ADRAN YMCHWIL

I'W DDEFNYDDIO YN Y LLYFRGELL
TO BE USED IN THE LIBRARY

The approved docum

corcolwyl

LL T4

What is an approved document?

The Secretary of State has approved a series of documents that give practical guidance about how to meet the requirements of the Building Regulations 2010 for England. These approved documents give guidance on each of the technical parts of the regulations and on regulation 7 (see the back of this document). The approved documents provide guidance for common building situations.

It is the responsibility of those carrying out building works to meet the requirements of the Buildings Regulations 2010. Although it is ultimately for the courts to determine whether those requirements have been met, the approved documents provide practical guidance on potential ways to achieve compliance with the requirements of the regulations in England.

Although approved documents cover common building situations, compliance with the guidance set out in the approved documents does not provide a guarantee of compliance with the requirements of the regulations because the approved documents cannot cater for all circumstances, variations and innovations. Those with responsibility for meeting the requirements of the regulations will need to consider for themselves whether following the guidance in the approved documents is likely to meet those requirements in the particular circumstances of their case.

Note that there may be other ways to comply with the requirements than the methods described in an approved document. If you prefer to meet a relevant requirement in some other way than that described in an approved document, you should seek to agree this with the relevant building control body at an early stage.

Where the guidance in the approved document has been followed, a court or inspector will tend to find that there is no breach of the regulations. However, where the guidance in the approved document has not been followed, this may be relied upon as tending to establish breach of the regulations and, in such circumstances, the person carrying out building works should demonstrate that the requirements of the regulations have been complied with by some other acceptable means or method.

In addition to guidance, some approved documents include provisions that must be followed exactly, as required by regulations or where methods of test or calculation have been prescribed by the Secretary of State.

Each approved document relates only to the particular requirements of the Building Regulations 2010 that the document addresses. However, building work must also comply with all other applicable requirements of the Building Regulations 2010 and all other applicable legislation.

How to use this approved document

This document uses the following conventions.

a. Text against a green background is an extract from the Building Regulations 2010 or the Building (Approved Inspectors etc.) Regulations 2010 (both as amended). These extracts set out the legal requirements of the regulations.

b. Key terms, printed in green, are defined in Appendix A.

GW 3526870 0

c. References are made to appropriate standards or other documents, which can provide further useful guidance. When this approved document refers to a named standard or other reference document, the standard or reference document has been clearly identified in this document. Standards are highlighted in **bold** throughout. The full name and version of the document referred to is listed in Appendix F (standards) or Appendix G (other documents). However, if the issuing body has revised or updated the listed version of the standard or document, you may use the new version as guidance if it continues to address the relevant requirements of the Building Regulations.

d. Standards and technical approvals also address aspects of performance or matters that are not covered by the Building Regulations and may recommend higher standards than required by the Building Regulations. Nothing in this approved document precludes you from adopting higher standards.

User requirements

The approved documents provide technical guidance. Users of the approved documents should have adequate knowledge and skills to understand and apply the guidance correctly to the building work being undertaken.

Where you can get further help

If you are not confident that you possess adequate knowledge and skills to apply the guidance correctly or if you do not understand the technical guidance or other information in this approved document or the additional detailed technical references to which it directs you, you should seek further help. Help can be obtained through a number of routes, some of which are listed below.

a. If you are the person undertaking the building work: either from your local authority building control service or from an approved inspector.

b. If you are registered with a competent person scheme: from the scheme operator.

c. If your query is technical: from a specialist or an industry technical body for the relevant subject.

The Building Regulations

The following is a high level summary of the Building Regulations relevant to most types of building work. Where there is any doubt you should consult the full text of the regulations, available at www.legislation.gov.uk.

Building work

Regulation 3 of the Building Regulations defines 'building work'. Building work includes:

a. the erection or extension of a building

b. the provision or extension of a controlled service or fitting

c. the material alteration of a building or a controlled service or fitting.

Regulation 4 states that building work should be carried out in such a way that, when work is complete:

a. For new buildings or work on a building that complied with the applicable requirements of the Building Regulations: the building complies with the applicable requirements of the Building Regulations.

b. For work on an existing building that did not comply with the applicable requirements of the Building Regulations:

   i. the work itself must comply with the applicable requirements of the Building Regulations, and

   ii. the building must be no more unsatisfactory in relation to the requirements than before the work was carried out.

Material change of use

Regulation 5 defines a 'material change of use' in which a building or part of a building that was previously used for one purpose will be used for another.

The Building Regulations set out requirements that must be met before a building can be used for a new purpose. To meet the requirements, the building may need to be altered in some way.

Materials and workmanship

In accordance with regulation 7, building work must be carried out in a workmanlike manner using adequate and proper materials. Guidance on regulation 7(1) is given in Approved Document 7 and guidance on regulation 7(2) is provided in Approved Document B.

Independent third party certification and accreditation

Independent schemes of certification and accreditation of installers can provide confidence that the required level of performance for a system, product, component or structure can be achieved.

Building control bodies may accept certification under such schemes as evidence of compliance with a relevant standard. However, a building control body should establish before the start of the building work that a scheme is adequate for the purposes of the Building Regulations.

Energy efficiency requirements

Part 6 of the Building Regulations imposes additional specific requirements for energy efficiency.

If a building is extended or renovated, the energy efficiency of the existing building or part of it may need to be upgraded.

Notification of work

Most building work and material changes of use must be notified to a building control body unless one of the following applies.

a. It is work that will be self-certified by a registered competent person or certified by a registered third party.

b. It is work exempted from the need to notify by regulation 12(6A) of, or Schedule 4 to, the Building Regulations.

Responsibility for compliance

People who are responsible for building work (e.g. agent, designer, builder or installer) must ensure that the work complies with all applicable requirements of the Building Regulations. The building owner may also be responsible for ensuring that work complies with the Building Regulations. If building work does not comply with the Building Regulations, the building owner may be served with an enforcement notice.

Contents

Section 0: Approved Document B: Fire safety – dwellings

Summary

0.1 This approved document has been published in two volumes. Volume 1 deals solely with dwellings, including blocks of flats, while Volume 2 deals with all other types of building covered by the Building Regulations.

Arrangement of sections

0.2 Requirements B1–B5 of Schedule 1 to the Building Regulations are dealt with separately in one or more sections. Each requirement is shown at the start of the relevant sections.

0.3 The provisions in this document have the following aims.

Requirement B1: When there is a fire, ensure both:

a. satisfactory means of sounding an alarm

b. satisfactory means of escape for people.

Requirement B2: Inhibit the spread of fire over internal linings of buildings.

Requirement B3: The building must be built such that all of the following are achieved in the event of a fire:

a. the premature collapse of the building is avoided

b. sufficient fire separation is provided within buildings and between adjoining buildings

c. automatic fire suppression is provided where necessary

d. the unseen spread of fire and smoke in cavities is restricted.

Requirement B4: Restrict both:

a. the potential for fire to spread over external walls and roofs (including compliance with regulations 6(4) and 7(2))

b. the spread of fire from one building to another.

Requirement B5: Ensure both:

a. satisfactory access for the fire service and its appliances

b. facilities in buildings to help firefighters save the lives of people in and around buildings.

Regulation 38: Provide fire safety information to building owners.

0.4 Guidance is given on each aspect separately, though many are closely interlinked. The document should be considered as a whole. The relationship between different requirements and their interdependency should be recognised. Particular attention should be given to the situation where one part of the guidance is not fully followed, as this could have a negative effect on other provisions.

Appendices: Information common to more than one requirement of Part B

0.5 Guidance on matters that refer to more than one section of this document can be found in the following appendices.

Appendix A: Key terms

Appendix B: Performance of materials, products and structures

Appendix C: Fire doorsets

Appendix D: Methods of measurement

Appendix E: Sprinklers

Appendix F: Standards referred to

Appendix G: Documents referred to

Management of premises

0.6 The Building Regulations do not impose any requirements on the management of a building, but do assume that it will be properly managed. This includes, for example, keeping protected escape routes virtually 'fire sterile'.

Appropriate fire safety design considers the way in which a building will be managed. Any reliance on an unrealistic or unsustainable management regime cannot be considered to have met the requirements of the regulations.

Once the building is in use, the management regime should be maintained and a suitable risk assessment undertaken for any variation in that regime. Failure to take proper management responsibility may result in the prosecution of an employer, building owner or occupier under legislation such as the Regulatory Reform (Fire Safety) Order 2005.

Property protection

0.7 The Building Regulations are intended to ensure a reasonable standard of life safety in a fire. The protection of property, including the building itself, often requires additional measures. Insurers usually set higher standards before accepting the insurance risk.

Many insurers use the *RISCAuthority Design Guide for the Fire Protection of Buildings* by the Fire Protection Association (FPA) as a basis for providing guidance to the building designer on what they require.

Further information on the protection of property can be obtained from the FPA website: www.thefpa.co.uk.

Inclusive design

0.8 The fire safety aspects of the Building Regulations aim to achieve reasonable standards of health and safety for people in and around buildings.

People, regardless of ability, age or gender, should be able to access buildings and use their facilities. The fire safety measures incorporated into a building should take account of the needs of everyone who may access the building, both as visitors and as people who live or work in it. It is not appropriate, except in exceptional circumstances, to assume that certain groups of people will be excluded from a building because of its use.

The provisions in this approved document are considered to be of a reasonable standard for most buildings. However, some people's specific needs might not be addressed. In some situations, additional measures may be needed to accommodate these needs. This should be done on a case-by-case basis.

Alternative approaches

0.9 The fire safety requirements of the Building Regulations will probably be satisfied by following the relevant guidance in this approved document. However, approved documents provide guidance for some common building situations, and there may be alternative methods of complying with the Building Regulation requirements.

If alternative methods are adopted, the overall level of safety should not be lower than the approved document provides. It is the responsibility of those undertaking the work to demonstrate compliance.

If other standards or guidance documents are adopted, the relevant fire safety recommendations in those publications should be followed in their entirety. However, in some circumstances it may be necessary to use one publication to supplement another. Care must be taken when using supplementary guidance to ensure that an integrated approach is used in any one building.

Guidance documents intended specifically for assessing fire safety in existing buildings often include less onerous provisions than those for new buildings and are therefore unlikely to be appropriate for building work that is controlled by the Building Regulations.

Buildings for industrial and commercial activities that present a special fire hazard, e.g. those that sell fuels, may require additional fire precautions to those in this approved document.

Buildings of special architectural or historic interest

0.10 Where Part B applies to existing buildings, particularly buildings of special architectural or historic interest for which the guidance in this document might prove too restrictive, some variation of the provisions in this document may be appropriate. In such cases, it is appropriate to assess the hazard and risk in the particular case and consider a range of fire safety features in that context.

Sheltered housing

0.11 While many of the provisions in this approved document for means of escape from flats are applicable to sheltered housing, the nature of the occupancy may necessitate some additional fire protection measures. The extent of such measures will depend on the form of the development. For example, a group of specially adapted bungalows or two storey flats, with few communal facilities, will not need to be treated differently from other single storey or two storey dwellinghouses or flats.

Fire safety engineering

0.12 Fire safety engineering might provide an alternative approach to fire safety. Fire safety engineering may be the only practical way to achieve a satisfactory standard of fire safety in some complex buildings and in buildings that contain different uses.

Fire safety engineering may also be suitable for solving a specific problem with a design that otherwise follows the provisions in this document.

0.13 **BS 7974** and supporting published documents (PDs) provide a framework for and guidance on the application of fire safety engineering principles to the design of buildings.

Purpose groups

0.14 Building uses are classified within different purpose groups, which represent different levels of hazard (see Table 0.1). A purpose group can apply to a whole building or a compartment within the building, and should relate to the main use of the building or compartment.

0.15 Where a building or compartment has more than one use, it is appropriate to assign each different use to its own purpose group in the following situations.

    a.   If the ancillary use is a flat.

    b.   If both of the following apply.

         i.   The building or compartment has an area of more than $280m^2$.

         ii.   The ancillary use relates to an area that is more than one-fifth of the total floor area of the building or compartment.

    c.   In 'shop and commercial' (purpose group 4) buildings or compartments, if the ancillary use is storage and both of the following apply.

         i.   The building or compartment has an area of more than $280m^2$.

         ii.   The storage area comprises more than one-third of the total floor area of the building or compartment.

0.16 Where there are multiple main uses that are not ancillary to one another (for example, shops with independent offices above), each use should be assigned to a purpose group in its own right. Where there is doubt as to which purpose group is appropriate, the more onerous guidance should be applied.

0.17 In sheltered housing, the guidance in Approved Document B Volume 2 should be consulted for the design of communal facilities, such as a common lounge.

Table 0.1 Classification of purpose groups

Volume 1 purpose groups

Title	Group	Purpose for which the building or compartment of a building is intended to be used
Residential (dwellings)	1(a)[1]	Flat.
	1(b)[2]	Dwellinghouse that contains a habitable storey with a floor level a minimum of 4.5m above ground level up to a maximum of 18m.[3]
	1(c)[2][4]	Dwellinghouse that does not contain a habitable storey with a floor level a minimum of 4.5m above ground level.

Volume 2 purpose groups

Title	Group	Purpose

Residential (institutional)	2(a)	Hospital, home, school or other similar establishment, where people sleep on the premises. The building may be either of the following.
		• Living accommodation for, or accommodation for the treatment, care or maintenance of, either:
		  – people suffering from disabilities due to illness or old age or other physical or mental incapacity
		  – people under the age of 5 years.
		• A place of lawful detention.

Residential (other)	2(b)	Hotel, boarding house, residential college, hall of residence, hostel or any other residential purpose not described above.
Office	3	Offices or premises used for any of the following and their control: • administration • clerical work (including writing, bookkeeping, sorting papers, filing, typing, duplicating, machine calculating, drawing and the editorial preparation of matter for publication, police and fire and rescue service work) • handling money (including banking and building society work) • communications (including postal, telegraph and radio communications) • radio, television, film, audio or video recording • performance (premises not open to the public).

Shop and commercial	4	Shops or premises used for either of the following.
		• A retail trade or business (including selling food or drink to the public for immediate consumption, retail by auction, self-selection and over-the-counter wholesale trading, the business of lending books or periodicals for gain, the business of a barber or hairdresser, and the rental of storage space to the public).
		• Premises to which the public are invited either:
		  – to deliver or collect goods in connection with their hire, repair or other treatment
		  – (except in the case of repair of motor vehicles) where the public themselves may carry out such repairs or other treatments.

Table 0.1 Continued

Title	Group	Purpose for which the building or compartment of a building is intended to be used
Assembly and recreation	5	Place of assembly, entertainment or recreation, including any of the following: • bingo halls, broadcasting, recording and film studios open to the public, casinos, dance halls • entertainment, conference, exhibition and leisure centres • funfairs and amusement arcades • museums and art galleries, non-residential clubs, theatres, cinemas, concert halls • educational establishments, dancing schools, gymnasia, swimming pool buildings, riding schools, skating rinks, sports pavilions, sports stadia • law courts • churches and other buildings of worship, crematoria • libraries open to the public, non-residential day centres, clinics, health centres and surgeries • passenger stations and termini for air, rail, road or sea travel • public toilets • zoos and menageries.
Industrial	6	Factories and other premises used for any of the following: • manufacturing, altering, repairing, cleaning, washing, breaking up, adapting or processing any article • generating power • slaughtering livestock.
Storage and other non-residential[4]	7(a)	Either of the following: • place (other than described under 7(b)) for the storage or deposit of goods or materials • any building not within purpose groups 1 to 6.
	7(b)	Car parks designed to admit and accommodate only cars, motorcycles and passenger or light goods vehicles that weigh a maximum of 2500kg gross.

NOTES:

This table only applies to Part B.

See Approved Document B Volume 2 for guidance on buildings other than dwellings (purpose groups 2, 3, 4, 5, 6 and 7).

1. Includes live/work units that meet the provisions of paragraph 3.24.

2. Includes any surgeries, consulting rooms, offices or other accommodation that meets all of the following conditions.

    a. A maximum of 50m^2 in total.

    b. Part of a dwellinghouse.

    c. Used by an occupant of the dwellinghouse in a professional or business capacity.

3. Where very large (over 18m in height or with a 10m deep basement) or unusual dwellinghouses are proposed, some of the guidance for buildings other than dwellings may be needed.

4. All of the following are included in purpose group 1(c).

    a. A detached garage a maximum of 40m^2 in area.

    b. A detached open carport a maximum 40m^2 in area.

    c. A detached building that consists of a garage and open carport, each a maximum of 40m^2 in area.

Mixed use buildings

0.18 This approved document includes reference to selected guidance for buildings other than dwellings. For the design of mixed use buildings, Approved Document B Volume 2 should be consulted in addition to the guidance contained in this approved document.

0.19 Where a complex mix of uses exists, the effect that one use may have on another in terms of risk should be considered. It could be necessary to use guidance from both volumes, apply other guidance (such as from HTM 05-02 or *Building Bulletin 100*), and/or apply special measures to reduce the risk.

B1

Requirement B1: Means of warning and escape

These sections deal with the following requirement from Part B of Schedule 1 to the Building Regulations 2010.

Requirement	
Requirement	*Limits on application*
Means of warning and escape	
B1. The building shall be designed and constructed so that there are appropriate provisions for the early warning of fire, and appropriate means of escape in case of fire from the building to a place of safety outside the building capable of being safely and effectively used at all material times.	Requirement B1 does not apply to any prison provided under section 33 of the Prison Act 1952[a] (power to provide prisons, etc.).
	(a) 1952 c. 52; section 33 was amended by section 100 of the Criminal Justice and Public Order Act 1994 (c. 33) and by S.I. 1963/597.

Intention

In the Secretary of State's view, requirement B1 is met by achieving all of the following.

a. There are sufficient means for giving early warning of fire to people in the building.

b. All people can escape to a place of safety without external assistance.

c. Escape routes are suitably located, sufficient in number and of adequate capacity.

d. Where necessary, escape routes are sufficiently protected from the effects of fire and smoke.

e. Escape routes are adequately lit and exits are suitably signed.

f. There are appropriate provisions to limit the ingress of smoke to the escape routes, or to restrict the spread of fire and remove smoke.

g. For buildings containing flats, there are appropriate provisions to support a stay put evacuation strategy.

The extent to which any of these measures are necessary is dependent on the use of the building, its size and its height.

Building work and material changes of use subject to requirement B1 include both new and existing buildings.

Section 1: Fire detection and alarm systems

General provisions

1.1 All dwellings should have a fire detection and alarm system, minimum Grade D2 Category LD3 standard, in accordance with the relevant recommendations of **BS 5839-6**.

A higher standard of protection should be considered where occupants of a proposed dwelling would be at special risk from fire. Further advice on this is also given in **BS 5839-6**.

1.2 Smoke alarms should be mains operated and conform to **BS EN 14604**.

1.3 Heat alarms should be mains operated and conform to **BS 5446-2**.

1.4 Smoke and heat alarms should have a standby power supply, such as a battery (rechargeable or non-rechargeable) or capacitor. More information on power supplies is given in clause 15 of **BS 5839-6**.

NOTE: The term 'fire alarm system' describes the combination of components for giving an audible and/or other perceptible warning of fire.

NOTE: In this document, the term 'fire detection system' describes any type of automatic sensor network and associated control and indicating equipment. Sensors may be sensitive to smoke, heat, gaseous combustion products or radiation. Automatic sprinkler systems can also be used to operate a fire alarm system.

Large dwellinghouses

1.5 A large dwellinghouse has more than one storey, and at least one storey exceeds 200m^2.

1.6 A large dwellinghouse of two storeys (excluding basement storeys) should be fitted with a Grade A Category LD3 fire detection and alarm system, as described in **BS 5839-6**.

1.7 A large dwellinghouse of three or more storeys (excluding basement storeys) should be fitted with a Grade A Category LD2 fire detection and alarm system as described in **BS 5839-6**.

Extensions and material alterations

1.8 Where new habitable rooms are provided, a fire detection and alarm system should be installed where either of the following applies.

a. The room is provided above or below the ground storey.

b. The room is provided at the ground storey, without a final exit.

1.9 Smoke alarms should be provided in the circulation spaces of the dwelling in accordance with paragraphs 1.1 to 1.4.

NOTE: This does not apply where inner rooms are provided (see paragraph 2.11 for inner room requirements).

Blocks of flats

1.10 Each flat in a block should have alarms as set out in paragraphs 1.1 to 1.4. With effective compartmentation, a communal fire alarm system is not normally needed. In some buildings, detectors in common parts of the building may need to operate smoke control or other fire protection systems but do not usually sound an audible warning.

Student accommodation

1.11 In student residences that are designed and occupied as a block of flats, separate automatic detection should be provided in each self-contained flat where all of the following apply.

  a. A group of up to six students shares the flat.

  b. Each flat has its own entrance door.

  c. The compartmentation principles for flats in Section 7 have been followed.

Where a total evacuation strategy is adopted, the alarm system should follow the guidance for buildings other than dwellings in Volume 2 of Approved Document B.

Sheltered housing

1.12 The fire detection and alarm systems in flats should connect to a central monitoring point or alarm receiving centre. The systems should alert the warden or supervisor and identify the individual flat where a fire has been detected.

1.13 These provisions do not apply to the following.

  a. The common parts of a sheltered housing development, such as communal lounges.

  b. Sheltered accommodation in the 'residential (institutional)' or 'residential (other)' purpose groups (purpose group 2(a) or 2(b)).

In these parts, means of warning should follow the guidance for buildings other than dwellings in Volume 2 of Approved Document B.

Design and installation of systems

1.14 Fire detection and alarm systems must be properly designed, installed and maintained. A design, installation and commissioning certificate should be provided for fire detection and alarm systems. Third party certification schemes for fire protection products and related services are an effective means of providing assurances of quality, reliability and safety.

Interface between fire detection and alarm systems and other systems

1.15 Fire detection and alarm systems sometimes trigger other systems. The interface between systems must be reliable. Particular care should be taken if the interface is facilitated via another system. Where any part of **BS 7273** applies to the triggering of other systems, the recommendations of that part of **BS 7273** should be followed.

Section 2: Means of escape – dwellinghouses

Escape from the ground storey

2.1 See Diagram 2.1a. All habitable rooms (excluding kitchens) should have either of the following.

a. An opening directly onto a hall leading to a final exit.

b. An emergency escape window or door, as described in paragraph 2.10.

Escape from upper storeys a maximum of 4.5m above ground level

2.2 See Diagram 2.1b. Where served by only one stair, all habitable rooms (excluding kitchens) should have either of the following.

a. An emergency escape window or external door, as described in paragraph 2.10.

b. Direct access to a protected stairway, as described in paragraph 2.5a.

2.3 Two rooms may be served by a single window. A door between the rooms should provide access to the window without passing through the stair enclosure. Both rooms should have their own access to the internal stair.

Escape from upper storeys more than 4.5m above ground level

2.4 Dwellinghouses with one internal stair should comply with paragraphs 2.5 and 2.6. In dwellinghouses with more than one stair, the stairs should provide effective alternative means of escape. The stairs should be physically separated by either of the following.

a. Fire resisting construction (minimum REI 30).

b. More than one room.

Dwellinghouses with one storey more than 4.5m above ground level

2.5 See Diagram 2.1c. The dwellinghouse should have either of the following.

a. **Protected stairway** – a stair separated by fire resisting construction (minimum REI 30) at all storeys, that complies with one of the following.

   i. Extends to a final exit (Diagram 2.2a).

   ii. Gives access to a minimum of two ground level final exits that are separated from each other by fire resisting construction (minimum REI 30) and fire doorsets (minimum E 20) (Diagram 2.2b).

   Cavity barriers or a fire resisting ceiling (minimum EI 30) should be provided above a protected stairway enclosure (Diagram 2.3).

b. **Alternative escape route** – a top storey separated from lower storeys by fire resisting construction (minimum REI 30) and with an alternative escape route leading to its own final exit.

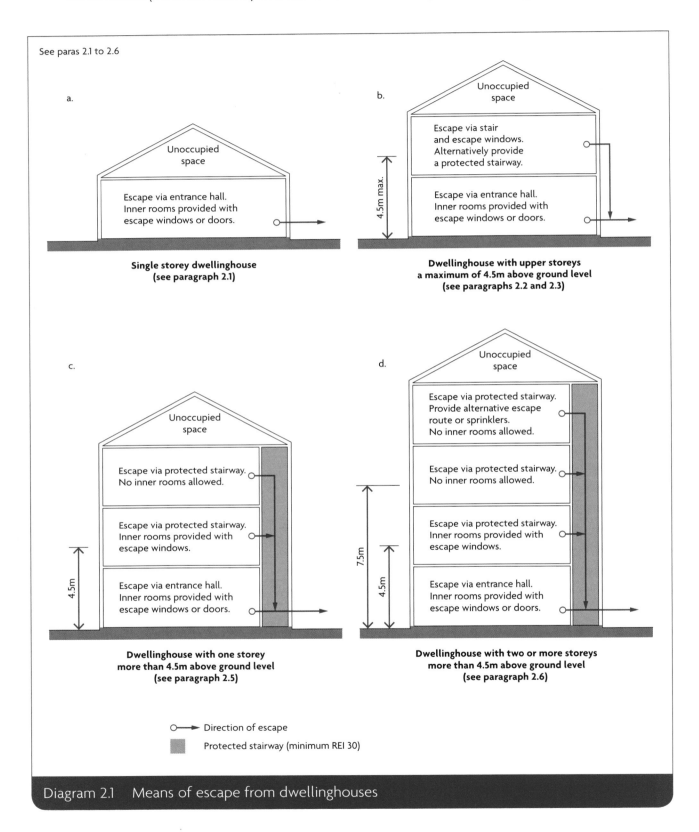

See paras 2.1 to 2.6

a.

Unoccupied space

Escape via entrance hall. Inner rooms provided with escape windows or doors.

Single storey dwellinghouse (see paragraph 2.1)

b.

Unoccupied space

Escape via stair and escape windows. Alternatively provide a protected stairway.

4.5m max.

Escape via entrance hall. Inner rooms provided with escape windows or doors.

Dwellinghouse with upper storeys a maximum of 4.5m above ground level (see paragraphs 2.2 and 2.3)

c.

Unoccupied space

Escape via protected stairway. No inner rooms allowed.

Escape via protected stairway. Inner rooms provided with escape windows.

4.5m

Escape via entrance hall. Inner rooms provided with escape windows or doors.

Dwellinghouse with one storey more than 4.5m above ground level (see paragraph 2.5)

d.

Unoccupied space

Escape via protected stairway. Provide alternative escape route or sprinklers. No inner rooms allowed.

Escape via protected stairway. No inner rooms allowed.

7.5m

Escape via protected stairway. Inner rooms provided with escape windows.

4.5m

Escape via entrance hall. Inner rooms provided with escape windows or doors.

Dwellinghouse with two or more storeys more than 4.5m above ground level (see paragraph 2.6)

○━━▶ Direction of escape

▭ Protected stairway (minimum REI 30)

Diagram 2.1   Means of escape from dwellinghouses

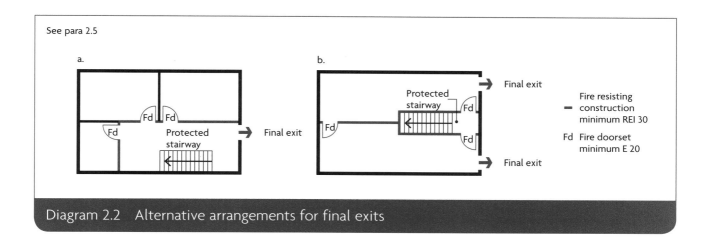

Diagram 2.2 Alternative arrangements for final exits

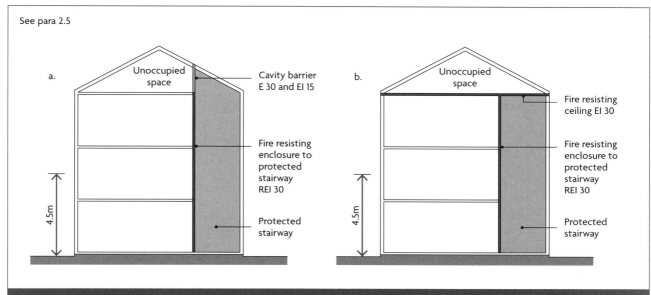

Diagram 2.3 Alternative cavity barrier arrangements in roof space over protected stairway in a house with a storey more than 4.5m above ground level

Dwellinghouses with two or more storeys more than 4.5m above ground level

2.6 See Diagram 2.1d. In addition to meeting the provisions in paragraph 2.5, the dwellinghouse should comply with either of the following.

a. Provide an alternative escape route from each storey more than 7.5m above ground level. At the first storey above 7.5m, the protected stairway should be separated from the lower storeys by fire resisting construction (minimum REI 30) if the alternative escape route is accessed via either of the following.

    i. The protected stairway to an upper storey.

    ii. A landing within the protected stairway enclosure to an alternative escape route on the same storey. The protected stairway at or about 7.5m above ground level should be separated from the lower storeys or levels by fire resisting construction (see Diagram 2.4).

b. Provide a sprinkler system throughout, designed and installed in accordance with **BS 9251**.

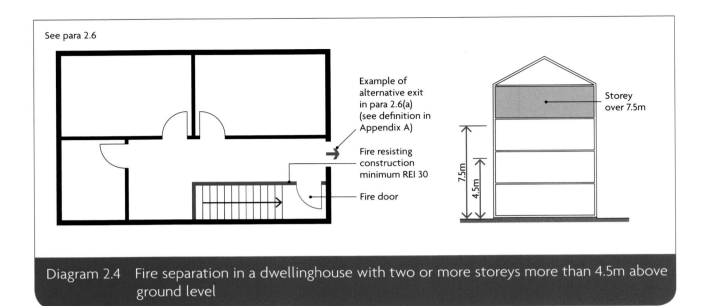

Diagram 2.4   Fire separation in a dwellinghouse with two or more storeys more than 4.5m above ground level

Passenger lifts

2.7   A passenger lift serving any storey more than 4.5m above ground level should be in either of the following.

   a.   The enclosure to the protected stairway, as described in paragraph 2.5.

   b.   A fire resisting lift shaft (minimum REI 30).

Air circulation systems

2.8   Air circulation systems which circulate air within an individual dwellinghouse with a floor more than 4.5m above ground level should meet the guidance given in paragraph 2.9.

2.9   All of the following precautions should be taken to avoid the spread of smoke and fire to the protected stairway.

   a.   Transfer grilles should not be fitted in any wall, door, floor or ceiling of the stair enclosure.

   b.   Any duct passing through the stair enclosure should be rigid steel. Joints between the ductwork and stair enclosure should be fire-stopped.

   c.   Ventilation ducts supplying or extracting air directly to or from a protected stairway should not serve other areas as well.

   d.   Any system of mechanical ventilation which recirculates air and which serves both the stair and other areas should be designed to shut down on the detection of smoke within the system.

   e.   For ducted warm air heating systems, a room thermostat should be sited in the living room. It should be mounted at a height between 1370mm and 1830mm above the floor. The maximum setting should be 27°C.

NOTE: Ventilation ducts passing through compartment walls should comply with the guidance in Section 9.

General provisions

Emergency escape windows and external doors

2.10 Windows or external doors providing emergency escape should comply with all of the following.

   a. Windows should have an unobstructed openable area that complies with all of the following.

      i.   A minimum area of 0.33m^2.

      ii. A minimum height of 450mm and a minimum width of 450mm (the route through the window may be at an angle rather than straight through).

      iii. The bottom of the openable area is a maximum of 1100mm above the floor.

   b. People escaping should be able to reach a place free from danger from fire. Courtyards or inaccessible back gardens should comply with Diagram 2.5.

   c. Locks (with or without removable keys) and opening stays (with child-resistant release catches) may be fitted to escape windows.

   d. Windows should be capable of remaining open without being held.

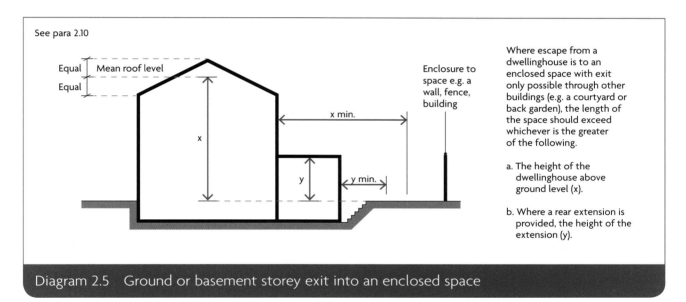

See para 2.10

Equal | Mean roof level
Equal

x

x min.

y

y min.

Enclosure to space e.g. a wall, fence, building

Where escape from a dwellinghouse is to an enclosed space with exit only possible through other buildings (e.g. a courtyard or back garden), the length of the space should exceed whichever is the greater of the following.

a. The height of the dwellinghouse above ground level (x).

b. Where a rear extension is provided, the height of the extension (y).

Diagram 2.5   Ground or basement storey exit into an enclosed space

Inner rooms

2.11 An inner room is permitted when it is one of the following.

   a. A kitchen.

   b. A laundry or utility room.

   c. A dressing room.

   d. A bathroom, WC or shower room.

   e. Any room on a storey that is a maximum of 4.5m above ground level which is provided with an emergency escape window as described in paragraph 2.10.

   f. A gallery that complies with paragraph 2.15.

2.12 A room accessed only via an inner room (an inner inner room) is acceptable when all of the following apply.

    a. It complies with paragraph 2.11.

    b. The access rooms each have a smoke alarm (see Section 1).

    c. None of the access rooms is a kitchen.

Balconies and flat roofs

2.13 Where a flat roof forms part of a means of escape, it should comply with all of the following.

    a. It should be part of the same building from which escape is being made.

    b. The route across the roof should lead to a storey exit or external escape route.

    c. The part of the roof (including its supporting structure) forming the escape route, and any opening within 3m of the escape route, should be of fire resisting construction (minimum REI 30).

2.14 A balcony or flat roof intended to form part of an escape route should be provided with guarding etc. in accordance with Approved Document K.

Galleries

2.15 A gallery should comply with one of the following.

    a. It should be provided with an alternative exit.

    b. It should be provided with an emergency escape window, as described in paragraph 2.10, where the gallery floor is a maximum of 4.5m above ground level.

    c. It should meet all the conditions shown in Diagram 2.6.

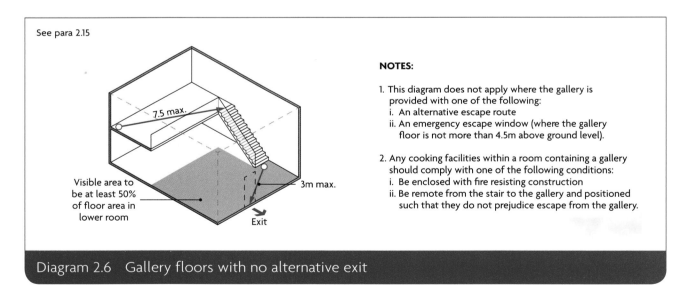

Diagram 2.6   Gallery floors with no alternative exit

Basements

2.16 Basement storeys containing habitable rooms should have one of the following.

    a. An emergency escape window or external door providing escape from the basement (paragraph 2.10).

    b. A protected stairway (paragraph 2.5a) leading from the basement to a final exit.

External escape stairs

2.17 Any external escape stair should meet all of the following conditions (Diagram 2.7).

a. Doors to the stair should be fire resisting (minimum E 30), except for a single exit door from the building to the top landing of a downward-leading external stair.

b. Fire resisting construction (minimum RE 30) is required for the building envelope within the following zones, measured from the flights and landings of the external stair.

   i. 1800mm horizontally.

   ii. 9m vertically below.

   iii. 1100mm above the top landing of the stair (except where the stair leads from basement to ground level).

c. Fire resisting construction (minimum RE 30) should be provided for any part of the building (including doors) within 1800mm of the escape route from the foot of the stair to a place of safety. This does not apply if there are alternative escape routes from the foot of the external escape stair.

d. Stairs more than 6m in height should be protected from adverse weather. Protection should prevent the build-up of snow or ice but does not require full enclosure.

e. Glazing in areas of fire resisting construction should be fixed shut and fire resisting (in terms of integrity, but not insulation) (minimum E 30).

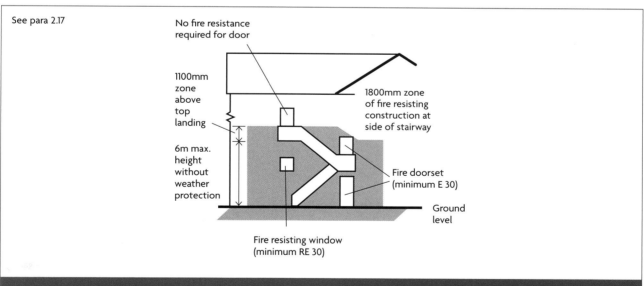

See para 2.17

No fire resistance required for door

1100mm zone above top landing

1800mm zone of fire resisting construction at side of stairway

6m max. height without weather protection

Fire doorset (minimum E 30)

Ground level

Fire resisting window (minimum RE 30)

Diagram 2.7   Fire resistance of areas near to external stairs

Work on existing dwellinghouses

Replacement windows

2.18 Work should comply with Parts K and L of Schedule 1 to the Building Regulations. When complete, the building should comply with other applicable parts of Schedule 1 to at least the same level as before.

2.19 Where an existing window would be an escape window in a new dwellinghouse, and is big enough to be used for escape purposes, then the replacement should comply with one of the following.

    a. The replacement window should be sized to provide at least the same potential for escape.

    b. If the existing window was larger than required for escape purposes, the opening can be reduced to the minimum described in paragraph 2.10.

2.20 If windows are replaced, it may be necessary to provide cavity barriers around the opening in accordance with Section 5.

Loft conversions

2.21 Where a new storey is added through conversion to create a storey above 4.5m, both of the following should apply.

    a. The full extent of the escape route should be addressed.

    b. Fire resisting doors (minimum E 20) and partitions (minimum REI 30) should be provided, including upgrading the existing doors where necessary.

    NOTE: Where the layout is open plan, new partitions should be provided to enclose the escape route (Diagram 2.2).

2.22 Where it is undesirable to replace existing doors because of historical or architectural merit, the possibility of retaining, and where necessary upgrading, them should be investigated.

2.23 An alternative approach to that described in paragraph 2.21 would be to comply with all of the following.

    a. Provide sprinkler protection to the open-plan areas.

    b. Provide a fire resisting partition (minimum REI 30) and door (minimum E 20) to separate the ground storey from the upper storeys. The door should allow occupants of the loft room access to a first storey escape window.

    c. Separate cooking facilities from the open-plan area with fire resisting construction (minimum REI 30).

Section 3: Means of escape – flats

Introduction

3.1 Separate guidance applies to means of escape within the flat and within the common parts of the building that lead to a place of safety. Flats at ground level are treated similarly to dwellinghouses. With increasing height, more complex provisions are needed.

3.2 The provisions in this section make the following assumptions.

    a. Any fire is likely to be in a flat.

    b. There is no reliance on external rescue.

    c. Simultaneous evacuation of all flats is unlikely to be necessary due to compartmentation.

    d. Fires in common parts of the building should not spread beyond the fabric in the immediate vicinity. In some cases, however, communal facilities exist that require additional measures to be taken.

3.3 Provisions are recommended to support a stay put evacuation strategy for blocks of flats. It is based on the principle that a fire is contained in the flat of origin and common escape routes are maintained relatively free from smoke and heat. It allows occupants, some of whom may require assistance to escape in the event of a fire, in other flats that are not affected to remain.

Sufficient protection to common means of escape is necessary to allow occupants to escape should they choose to do so or are instructed/aided to by the fire service. A higher standard of protection is therefore needed to ensure common escape routes remain available for a longer period than is provided in other buildings.

3.4 Paragraphs 3.6 to 3.23 deal with the means of escape within each flat. Paragraphs 3.25 to 3.89 deal with the means of escape in common areas of the building (including mixed use buildings in paragraphs 3.76 and 3.77). Guidance for live/work units is given in paragraph 3.24.

General provisions

Mixed use buildings

3.5 In mixed use buildings, separate means of escape should be provided from any storeys or parts of storeys used for the 'residential' or 'assembly and recreation' purpose groups (purpose groups 1, 2 and 5), other than in the case of certain small buildings or buildings in which the residential accommodation is ancillary (see paragraphs 3.76 and 3.77)

Emergency escape windows and external doors

3.6 Windows or external doors providing emergency escape should comply with all of the following.

    a. Windows should have an unobstructed openable area that complies with all of the following.

        i. A minimum area of 0.33m^2.

    ii.   A minimum height of 450mm and a minimum width of 450mm (the route through the window may be at an angle rather than straight through).

    iii.   The bottom of the openable area is a maximum of 1100mm above the floor.

  b. People escaping should be able to reach a place free from danger from fire.

  c. Locks (with or without removable keys) and opening stays (with child-resistant release catches) may be fitted to escape windows.

  d. Windows should be capable of remaining open without being held.

Inner rooms

3.7 An inner room is permitted when it is one of the following.

  a. A kitchen.

  b. A laundry or utility room.

  c. A dressing room.

  d. A bathroom, WC or shower room.

  e. Any room on a storey that is a maximum of 4.5m above ground level which is provided with an emergency escape window as described in paragraph 3.6.

  f. A gallery that complies with paragraph 3.13.

3.8 A room accessed only via an inner room (an inner inner room) is acceptable when all of the following apply.

  a. It complies with paragraph 3.7.

  b. The access rooms each have a smoke alarm (see Section 1).

  c. None of the access rooms is a kitchen.

Basements

3.9 Basement storeys containing habitable rooms should have one of the following.

  a. An emergency escape window or external door providing escape from the basement (see paragraph 3.6).

  b. A protected stairway (minimum REI 30) leading from the basement to a final exit.

Balconies and flat roofs

3.10 Where a flat roof forms part of a means of escape, it should comply with all of the following.

  a. It should be part of the same building from which escape is being made.

  b. The route across the roof should lead to a storey exit or external escape route.

  c. The part of the roof (including its supporting structure) forming the escape route, and any opening within 3m of the escape route, should be of fire resisting construction (minimum REI 30).

3.11 A balcony or flat roof intended to form part of an escape route should be provided with guarding etc. in accordance with Approved Document K.

3.12 For flats more than 4.5m above ground level, a balcony outside an alternative exit should be a common balcony meeting the conditions described in paragraph 3.22.

Galleries

3.13 A gallery should comply with one of the following.

   a.   It should be provided with an alternative exit.

   b.   It should be provided with an emergency escape window, as described in paragraph 3.6, where the gallery floor is a maximum of 4.5m above ground level.

   c.   It should meet the conditions shown in Diagram 3.1.

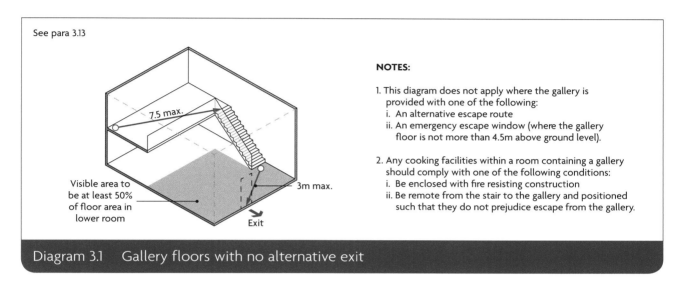

See para 3.13

7.5 max.

Visible area to be at least 50% of floor area in lower room

3m max.

Exit

NOTES:

1. This diagram does not apply where the gallery is provided with one of the following:
   i.   An alternative escape route
   ii. An emergency escape window (where the gallery floor is not more than 4.5m above ground level).

2. Any cooking facilities within a room containing a gallery should comply with one of the following conditions:
   i.   Be enclosed with fire resisting construction
   ii. Be remote from the stair to the gallery and positioned such that they do not prejudice escape from the gallery.

Diagram 3.1   Gallery floors with no alternative exit

Flats with upper storeys a maximum of 4.5m above ground level

3.14 The internal arrangement of single storey or multi-storey flats should comply with paragraphs 3.15 to 3.17. Alternatively, the guidance in paragraphs 3.18 to 3.22 may be followed.

A flat accessed via the common parts of the building should also comply with the provisions for small single stair buildings in paragraph 3.28 and Diagram 3.9. A protected entrance hall may be required as a result.

Escape from the ground storey

3.15 All habitable rooms (excluding kitchens) should have either of the following.

   a.   An opening directly onto a hall leading to a final exit.

   b.   An emergency escape window or door, as described in paragraph 3.6.

Escape from upper storeys a maximum of 4.5m above ground level

3.16 All habitable rooms (excluding kitchens) should have either of the following.

   a.   An emergency escape window or external door, as described in paragraph 3.6.

   b.   In multi-storey flats, direct access to a protected internal stairway (minimum REI 30) leading to an exit from the flat.

3.17 Two rooms may be served by a single escape window. A door between rooms should provide access to the escape window without passing through the stair enclosure. Both rooms should have their own access to the internal stair.

Flats with storeys more than 4.5m above ground level

Internal planning of single storey flats

3.18 One of the following approaches should be adopted, observing the inner room restrictions described in paragraphs 3.7 and 3.8.

a. Provide a protected entrance hall (minimum REI 30) serving all habitable rooms that meets the conditions shown in Diagram 3.2.

b. Plan the flat to meet the conditions shown in Diagram 3.3, so that both of the following apply.

   i. The travel distance from the flat entrance door to any point in any habitable room is a maximum of 9m.

   ii. Cooking facilities are remote from the main entrance door and do not impede the escape route from anywhere in the flat.

c. Provide an alternative exit from the flat complying with paragraph 3.22.

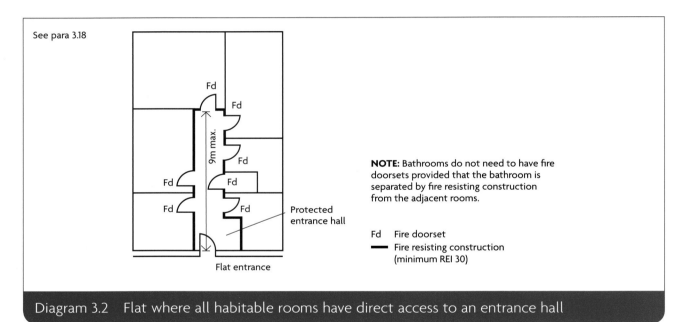

Diagram 3.2   Flat where all habitable rooms have direct access to an entrance hall

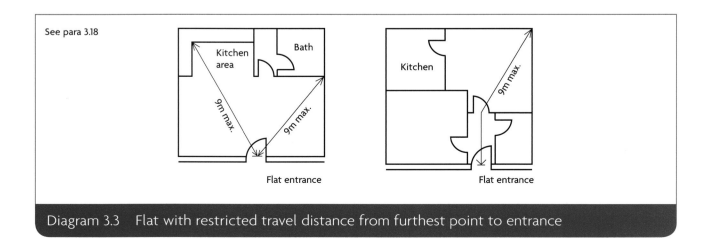

Diagram 3.3   Flat with restricted travel distance from furthest point to entrance

Flats with an alternative exit

3.19 Where access from any habitable room to the entrance hall or flat entrance is impossible without passing through another room, all of the following conditions should be met (Diagram 3.4).

    a. Bedrooms should be separated from living accommodation by fire resisting construction (minimum REI 30) and fire doorsets (minimum E 20).

    b. The alternative exit should be in the part of the flat that contains the bedrooms.

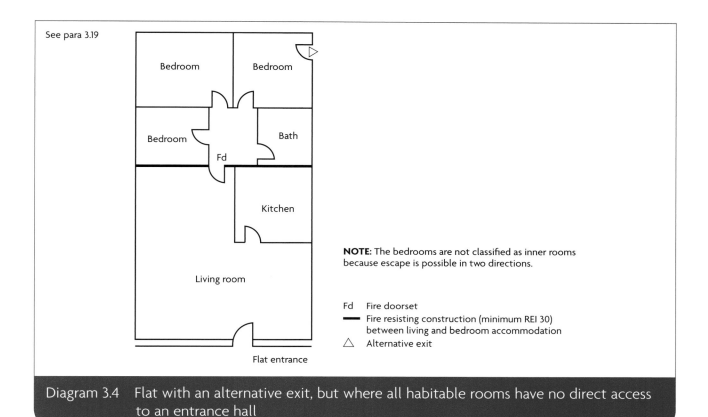

Diagram 3.4 Flat with an alternative exit, but where all habitable rooms have no direct access to an entrance hall

Internal planning of multi-storey flats

3.20 A multi-storey flat with an independent external entrance at ground level is similar to a dwellinghouse and means of escape should be planned on the basis of Section 2, depending on the height of the top storey above ground level.

3.21 When multi-storey flats do not have their own external entrance at ground level, adopt one of the following approaches.

    a. **Approach 1** – provide at least one alternative exit from each habitable room that is not on the entrance storey of the flat (Diagram 3.5 and paragraph 3.22).

    b. **Approach 2** – provide at least one alternative exit from each storey that is not the entrance storey of the flat. All habitable rooms should have direct access to a protected landing (Diagram 3.6 and paragraph 3.22).

    c. **Approach 3** – provide a protected stairway plus a sprinkler system in accordance with Appendix E and provide smoke alarms in accordance with **BS 5839-6**.

d. **Approach 4** – if the vertical distance between the entrance storey of the flat and any of the storeys above or below does not exceed 7.5m, provide all of the following.

  i. A protected stairway.

  ii. Additional smoke alarms in all habitable rooms.

  iii. A heat alarm in any kitchen.

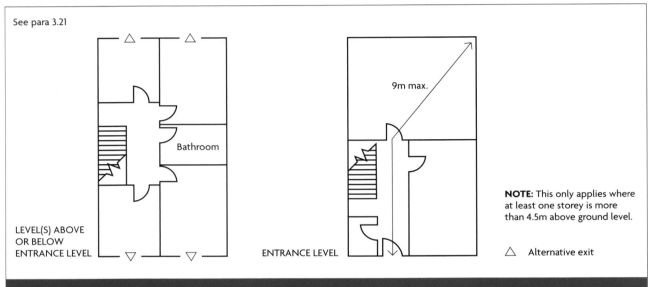

Diagram 3.5 Multi-storey flat with alternative exits from each habitable room, except at entrance level

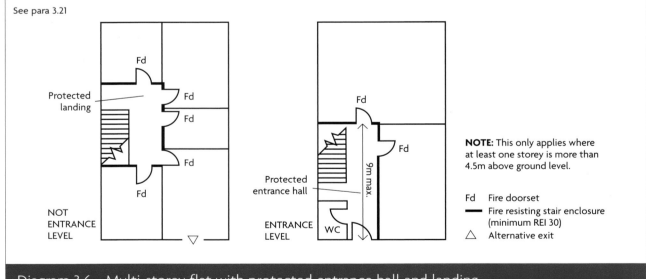

Diagram 3.6 Multi-storey flat with protected entrance hall and landing

Alternative exits

3.22 Any alternative exit from a flat should comply with all of the following.

a. It should be remote from the main entrance door to the flat.

b. It should lead to a final exit, via a common stair if necessary, through one of the following.

   i. A door to an access corridor, access lobby or common balcony.

   ii. An internal private stair leading to an access corridor, access lobby or common balcony at another level.

   iii. A door to a common stair.

   iv. A door to an external stair.

   v. A door to an escape route over a flat roof.

Any access route leading to a final exit or common stair should comply with the provisions for means of escape in the common parts of a flat (see paragraph 3.25).

Air circulation systems in flats with a protected stairway or entrance hall enclosure

3.23 For systems circulating air only within an individual flat, take all of the following precautions.

a. Transfer grilles should not be fitted in any wall, door, floor or ceiling of the enclosure.

b. Any duct passing through the enclosure should be rigid steel. Joints between the ductwork and enclosure should be fire-stopped.

c. Ventilation ducts serving the enclosure should not serve any other areas.

d. Any system of mechanical ventilation which recirculates air and which serves both the stair and other areas should be designed to shut down on the detection of smoke within the system.

e. For ducted warm air heating systems, a room thermostat should be sited in the living room. It should be mounted at a height between 1370mm and 1830mm above the floor. The maximum setting should be 27°C.

NOTE: Ventilation ducts passing through compartment walls should comply with the guidance in Section 9.

Live/work units

3.24 For flats serving as a workplace for both occupants and people who do not live on the premises, provide both of the following.

a. A maximum travel distance of 18m between any part of the working area and either of the following.

   i. The flat entrance door.

   ii. An alternative means of escape that is not a window.

If the travel distance is over 18m, the assumptions in paragraph 3.2 may not be valid. The design should be considered on a case-by-case basis.

b. Escape lighting to windowless accommodation in accordance with **BS 5266-1**.

Means of escape in the common parts of flats

3.25 The following paragraphs deal with means of escape from the entrance doors of flats to a final exit. They do not apply to flats with a top storey that is a maximum of 4.5m above ground level (designed in accordance with paragraphs 3.16 and 3.17).

Reference should also be made to the following.

a. Requirement B3 regarding compartment walls and protected shafts.

b. Requirement B5 regarding access for the fire and rescue service.

Number of escape routes

3.26 A person escaping through the common area, if confronted by the effects of a fire in another flat, should be able to turn away from it and make a safe escape via an alternative route.

3.27 From the flat entrance door, a single escape route is acceptable in either of the following cases.

a. The flat is on a storey served by a single common stair and both of the following apply.

    i. Every flat is separated from the common stair by a protected lobby or common protected corridor (see Diagram 3.7).

    ii. The maximum travel distance in Table 3.1, for escape in one direction only, is not exceeded.

b. The flat is in a dead end of a common corridor served by two (or more) common stairs and the maximum travel distance given in Table 3.1, for escape in one direction only, is not exceeded (Diagram 3.8).

Table 3.1 Limitations on travel distance in common areas of blocks of flats	
Maximum travel distance from flat entrance door to common stair or stair lobby[1]	
Escape in one direction	Escape in more than one direction
7.5m[2][3]	30m[3][4]

NOTES:

1. If travel distance is measured to a stair lobby, the lobby must not provide direct access to any storage room, flat or other space containing a fire hazard.

2. In the case of a small single stair building in accordance with Diagram 3.9, this is reduced to 4.5m.

3. Does not apply if all flats on a storey have independent alternative means of escape.

4. Sheltered housing may require reduced maximum travel distances.

See paras 3.27 and 3.36

a. CORRIDOR ACCESS FLATS

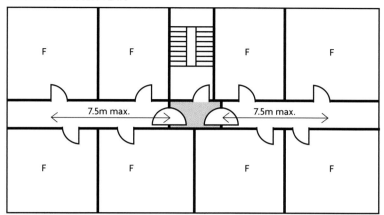

b. LOBBY ACCESS FLATS

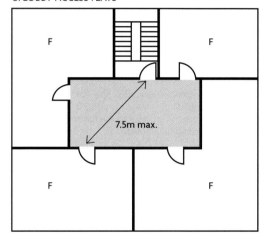

NOTES:

1. The arrangements shown also apply to the top storey.
2. See Diagram 3.9 for small single stair buildings.
3. All doors shown are fire doorsets.
4. Where travel distance is measured to a stair lobby, the lobby must not provide direct access to any storage room, flat or other space containing a potential fire hazard.
5. For further guidance on the fire rating of the fire doorsets from the corridor to the flat and/or stairway refer to Appendix C, Table C1.

F Flat

▨ Shaded areas indicate zones where ventilation should be provided in accordance with paragraphs 3.50 to 3.53 (An external wall vent or smoke shaft located anywhere in the shaded area)

Diagram 3.7   Flats served by one common stair

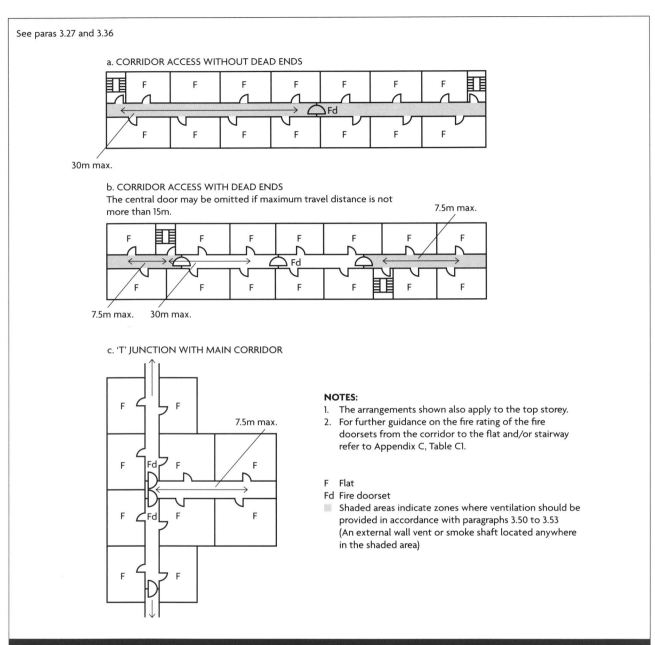

See paras 3.27 and 3.36

a. CORRIDOR ACCESS WITHOUT DEAD ENDS

30m max.

b. CORRIDOR ACCESS WITH DEAD ENDS
The central door may be omitted if maximum travel distance is not more than 15m.

7.5m max.

7.5m max.   30m max.

c. 'T' JUNCTION WITH MAIN CORRIDOR

7.5m max.

NOTES:
1. The arrangements shown also apply to the top storey.
2. For further guidance on the fire rating of the fire doorsets from the corridor to the flat and/or stairway refer to Appendix C, Table C1.

F   Flat
Fd   Fire doorset
Shaded areas indicate zones where ventilation should be provided in accordance with paragraphs 3.50 to 3.53 (An external wall vent or smoke shaft located anywhere in the shaded area)

Diagram 3.8   Flats served by more than one common stair

Small single stair buildings

3.28 For some low rise buildings, the provisions in paragraphs 3.26 and 3.27 may be modified and the use of a single stair, protected in accordance with Diagram 3.9, may be permitted where all of the following apply.

a. The top storey of the building is a maximum of 11m above ground level.

b. No more than three storeys are above the ground storey.

c. The stair does not connect to a covered car park, unless the car park is open sided (as defined in Section 11 of Approved Document B Volume 2).

d. The stair does not serve offices, stores or other ancillary accommodation. If it does, they should be separated from the stair by a protected lobby or protected corridor (minimum REI 30) with a minimum 0.4m² of permanent ventilation, or be protected from the ingress of smoke by a mechanical smoke control system.

NOTE: For refuse chutes and storage see paragraphs 3.55 to 3.58.

e. Either of the following is provided for the fire and rescue service.

  i. A high-level openable vent with a free area of at least 1m² at each storey.

  ii. A single openable vent with a free area of at least 1m² at the head of the stair, operable remotely at the fire and rescue service access level.

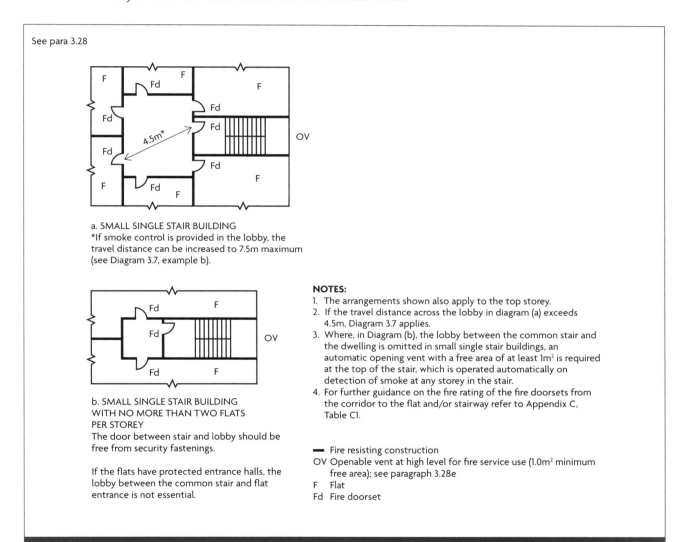

See para 3.28

a. SMALL SINGLE STAIR BUILDING
*If smoke control is provided in the lobby, the travel distance can be increased to 7.5m maximum (see Diagram 3.7, example b).

b. SMALL SINGLE STAIR BUILDING WITH NO MORE THAN TWO FLATS PER STOREY
The door between stair and lobby should be free from security fastenings.

If the flats have protected entrance halls, the lobby between the common stair and flat entrance is not essential.

NOTES:
1. The arrangements shown also apply to the top storey.
2. If the travel distance across the lobby in diagram (a) exceeds 4.5m, Diagram 3.7 applies.
3. Where, in Diagram (b), the lobby between the common stair and the dwelling is omitted in small single stair buildings, an automatic opening vent with a free area of at least 1m² is required at the top of the stair, which is operated automatically on detection of smoke at any storey in the stair.
4. For further guidance on the fire rating of the fire doorsets from the corridor to the flat and/or stairway refer to Appendix C, Table C1.

━ Fire resisting construction
OV Openable vent at high level for fire service use (1.0m² minimum free area); see paragraph 3.28e
F   Flat
Fd Fire doorset

Diagram 3.9   Common escape route in small single stair building

Flats with balcony or deck access

3.29 Paragraph 3.27 may be modified using the guidance in clause 7.3 of **BS 9991**.

Escape routes over flat roofs

3.30 Where a storey or part of a building has multiple escape routes available, one may be over a flat roof that complies with all of the following.

a. It should be part of the same building from which escape is being made.

b. The route across the roof should lead to a storey exit or external escape route.

c. The part of the roof (including its supporting structure) forming the escape route, and any opening within 3m of the escape route, should be of fire resisting construction (minimum REI 30).

d. The route should be clearly defined and guarded by walls and/or protective barriers to protect against falling.

Common escape routes

3.31 The following paragraphs deal with means of escape from the entrance doors of flats to a final exit.

3.32 Escape route travel distances should comply with Table 3.1.

3.33 An escape route should not pass through one stair enclosure to reach another. It may pass through a protected lobby (minimum REI 30) of one stair to reach another.

3.34 Common corridors should be protected corridors. The wall between each flat and the corridor should be a compartment wall (minimum REI 30 where the top storey is up to 5m above ground level, otherwise REI 60).

3.35 Divide a common corridor connecting two or more storey exits with a fire doorset fitted with a self-closing device (minimum E 30 S_a). See Diagram 3.8. Associated screens should be fire resisting. Site doors so that smoke does not affect access to more than one stair.

3.36 A fire doorset (minimum E 30 S_a) fitted with a self-closing device (and fire resisting screen, where required) should separate the dead-end portion of a common corridor from the rest of the corridor (Diagrams 3.7a, 3.8b and 3.8c).

3.37 Ancillary accommodation should not be located in, or entered from, a protected lobby or protected corridor forming the only common escape route on that storey.

Headroom in common escape routes

3.38 Escape routes should have a minimum clear headroom of 2m. The only projections allowed below this height are door frames.

Flooring of common escape routes

3.39 Escape route floor finishes should minimise their slipperiness when wet. Finishes include the treads of steps and surfaces of ramps and landings.

Ramps and sloping floors

3.40 A ramp forming part of an escape route should meet the provisions in Approved Document M. Any sloping floor or tier should have a pitch of not more than 35 degrees to the horizontal.

Lighting of common escape routes

3.41 All escape routes should have adequate artificial lighting. If the mains electricity power supply fails, escape lighting should illuminate the route (including external escape routes).

3.42 In addition, escape lighting should be provided to all of the following.

    a.   Toilet accommodation with a minimum floor area of 8m².

    b.   Electricity and generator rooms.

    c.   Switch room/battery room for emergency lighting system.

    d.   Emergency control rooms.

3.43 Escape stair lighting should be on a separate circuit from the electricity supply to any other part of the escape route.

3.44 Escape lighting should conform to **BS 5266-1**.

Exit signs on common escape routes

3.45 Every doorway or other exit providing access to a means of escape, other than exits in ordinary use (e.g. main entrances), should be distinctively and conspicuously marked by an exit sign in accordance with **BS ISO 3864-1** and **BS 5499-4**. For this reason, blocks of flats with a single stair in regular use would not usually require any fire exit signage.

Advice on fire safety signs, including emergency escape signs, is given in the HSE publication *Safety Signs and Signals: Guidance on Regulations*.

Some buildings may require additional signs to comply with other legislation.

Protected power circuits

3.46 To limit potential damage to cables in protected circuits, all of the following should apply.

    a.   Cables should be sufficiently robust.

    b.   Cable routes should be carefully selected and/or physically protected in areas where cables may be exposed to damage.

    c.   Methods of cable support should be class A1 rated and offer at least the same integrity as the cable. They should maintain circuit integrity and hold cables in place when exposed to fire.

3.47 A protected circuit to operate equipment during a fire should achieve all of the following.

    a.   Cables should achieve PH 30 classification when tested in accordance with **BS EN 50200** (incorporating Annex E) or an equivalent standard.

    b.   It should only pass through parts of the building in which the fire risk is negligible.

    c.   It should be separate from any circuit provided for another purpose.

3.48 Guidance on cables for large and complex buildings is given in **BS 5839-1**, **BS 5266-1** and **BS 8519**.

Smoke control in common escape routes

3.49 Despite the provisions described, it is probable that some smoke will get into the common corridor or lobby from a fire in a flat.

There should therefore be some means of ventilating the common corridors/lobbies to control smoke and so protect the common stairs. This means of ventilation offers additional protection to that provided by the fire doors to the stair, as well as some protection to the corridors/lobbies.

Ventilation can be natural (paragraphs 3.50 to 3.53) or mechanical (paragraph 3.54).

Smoke control of common escape routes by natural smoke ventilation

3.50 Except in buildings that comply with Diagram 3.9, the corridor or lobby next to each stair should have a smoke vent. The location of the vent should comply with both of the following.

a. Be as high as practicable.

b. Be positioned so the top edge is at least as high as the top of the door to the stair.

3.51 Smoke vents should comply with one of the following.

a. They should be located on an external wall with minimum free area of $1.5m^2$.

b. They should discharge into a vertical smoke shaft, closed at the base, that meets all of the following criteria.

   i. The shaft should conform to the following conditions.

   - Have a minimum cross-sectional area of $1.5m^2$ (minimum dimension 0.85m in any direction).

   - Open at roof level, minimum 0.5m above any surrounding structures within 2m of it horizontally.

   - Extend a minimum of 2.5m above the ceiling of the highest storey served by the shaft.

   ii. The free area of all the following vents should be a minimum of $1m^2$ in the following places.

   - From the corridor or lobby into the shaft.

   - At the opening at the head of the shaft.

   - At all internal locations within the shaft (e.g. safety grilles).

   iii. The smoke shaft should be constructed from a class A1 material. All vents should either be a fire doorset (see Appendix C, Table C1, item 2.e for minimum fire resistance) or fitted with a smoke control damper achieving the same period of fire resistance and designed to operate as described below. The shaft should be vertical from base to head, with a maximum of 4m at a maximum inclined angle of 30 degrees.

   iv. If smoke is detected in the common corridor or lobby, both of the following should occur.

   - Simultaneous opening of vents on the storey where the fire is located, at the top of the smoke shaft and to the stair.

   - Vents from the corridors or lobbies on all other storeys should remain closed, even if smoke is subsequently detected on storeys other than where the fire is located.

3.52 A vent to the outside with a minimum free area of $1m^2$ should be provided from the top storey of the stair.

3.53 In single stair buildings, smoke vents on the storey where the fire is initiated, and the vent at the head of the stair, should be activated by smoke detectors in the common parts.

In buildings with more than one stair, smoke vents may be activated manually. The control system should open the vent at the head of the stair before, or at the same time as, the vent on the storey where the fire is located. Smoke detection is not required for ventilation purposes in this instance.

Smoke control of common escape routes by mechanical ventilation

3.54 Guidance on the design of smoke control systems that use pressure differentials is available in **BS EN 12101-6**.

Refuse chutes and storage

3.55 Refuse storage chambers, refuse chutes and refuse hoppers should be sited and constructed in accordance with **BS 5906**.

3.56 Refuse chutes and rooms for storing refuse should meet both of the following conditions.

a. Be separated from other parts of the building by fire resisting construction (minimum REI 30 in buildings with a top storey up to 5m above ground level; otherwise REI 60).

b. Not be situated within a protected stairway or protected lobby.

3.57 The approach to rooms containing refuse chutes or for storing refuse should comply with one of the following conditions.

a. Be directly from the open air.

b. Be through a protected lobby with a minimum of 0.2m² of permanent ventilation.

3.58 Access openings to refuse storage chambers should *not* be sited in the following areas.

a. Next to escape routes or final exits.

b. Near the windows of flats.

Common stairs

Number of common stairs

3.59 A building should provide access to more than one common stair if it does not meet the criteria for a single common stair (see paragraph 3.26 and 3.27).

Width of common stairs

3.60 A stair of acceptable width for everyday use will be sufficient for escape purposes. If it is also a firefighting stair, it should be at least 1100mm wide. The width is the clear width between the walls or balustrades. Any handrails and strings intruding into that width by a maximum of 100mm on each side may be ignored.

Protection of common stairs

3.61 Section 7 provides guidance on avoiding the spread of fire between storeys. For a stair that is also a firefighting stair, guidance in Section 15 should be followed.

Enclosure of common stairs

3.62 Every common stair should be a protected stairway, within a fire resisting enclosure (minimum REI 30).

External walls adjacent to protected stairways

3.63 With some configurations of external wall, a fire in one part of a building could subject the external wall of a protected stairway to heat (for example, where the two are adjacent at an internal angle in the façade, as shown in Diagram 3.10).

3.64 If a protected stairway projects beyond, is recessed from or is in an internal angle of the adjoining external wall of the building, then the minimum distance between an unprotected area of the building enclosure and an unprotected area of the stair enclosure should be 1800mm.

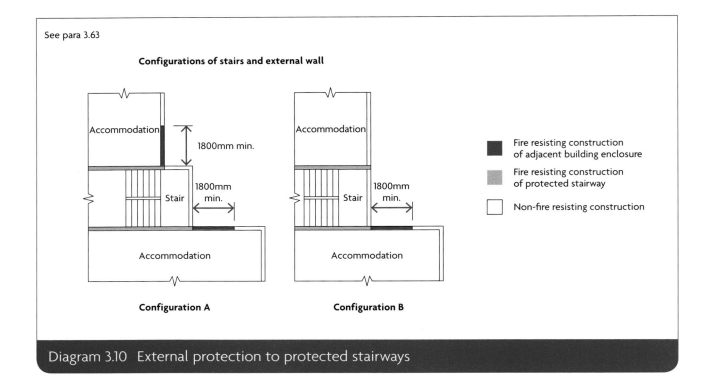

See para 3.63

Configurations of stairs and external wall

Accommodation

1800mm min.

Stair

1800mm min.

Accommodation

Configuration A

Accommodation

Stair

1800mm min.

Accommodation

Configuration B

■ Fire resisting construction of adjacent building enclosure

▤ Fire resisting construction of protected stairway

☐ Non-fire resisting construction

Diagram 3.10   External protection to protected stairways

External escape stairs

3.65 Flats may be served by an external stair if the provisions in paragraphs 3.66 to 3.69 are followed.

3.66 Where a storey (or part of a building) is served by a single access stair, that stair may be external provided both of the following conditions are met.

a.   The stair serves a floor not more than 6m above the ground level.

b.   The stair meets the provisions in paragraph 3.62.

3.67 Where more than one escape route is available from a storey (or part of a building), then some of the escape routes from that storey or part of the building may be by way of an external stair provided all of the following conditions are met:

a.   There is a at least one internal escape stair from every part of each storey (excluding plant areas).

b.   The stair serves a floor not more than 6m above either the ground level or a roof podium which is itself served by an independent protected stairway.

c.   The stair meets the provisions in paragraph 3.68.

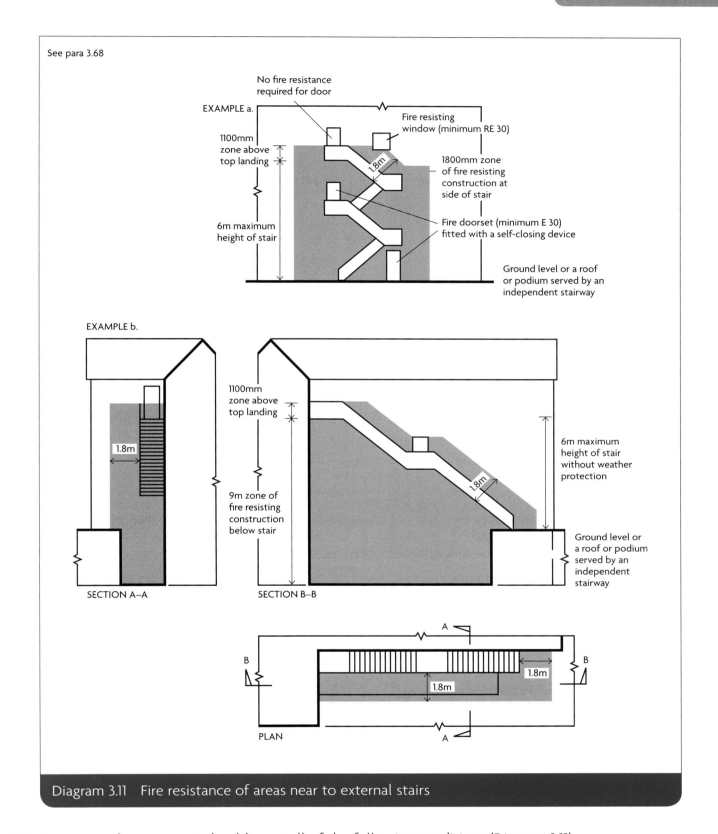

See para 3.68

EXAMPLE a.

No fire resistance required for door

1100mm zone above top landing

6m maximum height of stair

1.8m

Fire resisting window (minimum RE 30)

1800mm zone of fire resisting construction at side of stair

Fire doorset (minimum E 30) fitted with a self-closing device

Ground level or a roof or podium served by an independent stairway

EXAMPLE b.

1.8m

1100mm zone above top landing

9m zone of fire resisting construction below stair

1.8m

6m maximum height of stair without weather protection

Ground level or a roof or podium served by an independent stairway

SECTION A–A

SECTION B–B

A

B

B

1.8m

1.8m

PLAN

A

Diagram 3.11   Fire resistance of areas near to external stairs

3.68 Any external escape stair should meet all of the following conditions (Diagram 3.11).

a. Doors to the stair should be fire resisting (minimum E 30) and be fitted with a self-closing device, except for a single exit door from the building to the top landing of a downward-leading external stair, provided it is the only door onto the landing.

b. Fire resisting construction (minimum RE 30) is required for the building envelope within the following zones, measured from the flights and landings of the external stair.

   i. 1800mm above and horizontally.

   ii. 9m vertically below.

   iii. 1100mm above the top landing of the stair (except where the stair leads from basement to ground level).

c. Fire resisting construction (minimum RE 30) should be provided for any part of the building (including doors) within 1800mm of the escape route from the foot of the stair to a place of safety. This does not apply if there are alternative escape routes from the foot of the external escape stair.

d. Glazing in areas of fire resisting construction should be fixed shut and fire resisting (in terms of integrity but not insulation) (minimum E 30).

e. Stairs more than 6m in height above ground level (e.g. where they are provided above a podium) should be protected from adverse weather. Protection should prevent the build-up of snow or ice but does not require full enclosure.

3.69 Access to an external escape stair may be via a flat roof, provided the flat roof meets the requirements of paragraph 3.30.

Separation of adjoining protected stairways

3.70 The construction separating two adjacent protected stairways (or exit passageways leading to different final exits) should be imperforate.

Basement stairs

3.71 If a building does not meet the criteria of paragraph 3.28, an escape stair forming part of the only escape route from an upper storey should *not* continue down to serve a basement storey. The basement storey should be served by a separate escape stair.

3.72 Where multiple escape stairs serve the upper storeys, only one needs to end at ground level. Other stairs may connect with the basement storeys if there is a protected lobby or a protected corridor between the stairs and accommodation at each basement level.

Stairs serving ancillary accommodation

3.73 Except in buildings described in paragraph 3.28, common stairs forming part of the only escape route from a flat should not serve any of the following.

a. Covered car park.

b. Boiler room.

c. Fuel storage space.

d. Other ancillary accommodation of similar fire risk.

3.74 Where a common stair is not part of the only escape route from a flat, it may also serve ancillary accommodation from which it is separated by a protected lobby or protected corridor (minimum REI 30).

3.75 Where a stair serves an enclosed car park or place of special fire hazard, the lobby or corridor should have a minimum 0.4m² of permanent ventilation or be protected from the ingress of smoke by a mechanical smoke control system.

NOTE: For refuse chutes and storage see paragraphs 3.55 to 3.58.

Flats in mixed use buildings

3.76 In buildings with a maximum of three storeys above the ground storey, stairs may serve both flats and other occupancies, provided that the stairs are separated from each occupancy by protected lobbies (minimum REI 30) at each storey.

3.77 In buildings with more than three storeys above the ground storey, stairs may serve the flats and other occupancies if all of the following apply.

   a. The flat is ancillary to the main use of the building.

   b. The flat has an independent alternative escape route.

   c. The stair is separated from occupancies on lower storeys by a protected lobby (minimum REI 30) at each of those storeys.

   d. The stair enclosure has at least the same standard of fire resistance as the structural elements of the building (see Appendix B, Table B4); if the stair is a firefighting stair, it should comply with the provisions in Section 15 (see also paragraph 3.60).

   e. Any automatic fire detection and alarm system fitted in the main part of the building also covers all flats.

   f. Any security measures in any parts of the building do not prevent escape at all material times.

Use of space within protected stairways

3.78 A protected stairway should not be used for anything else, except a lift well or electricity meters.

Electricity meter(s) in protected stairways

3.79 In single stair buildings, electricity meters should be in securely locked cupboards. Cupboards should be separated from the escape route by fire resisting construction.

Gas service and installation pipes in protected stairways

3.80 Gas service and installation pipes and meters should not be within a protected stairway, unless installed in accordance with the Pipelines Safety Regulations 1996 and the Gas Safety (Installation and Use) Regulations 1998.

Exits from protected stairways

3.81 Every protected stairway should lead to a final exit, either directly or via a protected exit passageway. Any protected exit corridor or stair should have the same standard of fire resistance and lobby protection as the stair it serves.

Construction of escape stairs

3.82 The flights and landings of escape stairs should be constructed of materials achieving class A2-s3, d2 or better in all of the following situations.

   a. If the escape stair is the only stair in a building with more than three storeys.

   b. If the escape stair is within a basement storey.

   c. If the escape stair serves any storey that has a floor level more than 18m above ground or access level.

   d. If the escape stair is an external escape stair, except where the stair connects the ground storey or ground level with a floor or flat roof a maximum of 6m above or below ground level.

   e. If the escape stair is a firefighting stair.

Materials achieving class B-s3, d2 or worse may be added to the top horizontal surface, except on firefighting stairs.

3.83 Further guidance on the construction of firefighting stairs is given in Section 15 (see also paragraph 3.60). Dimensional constraints on the design of stairs are given in Approved Document K.

Single steps

3.84 Single steps on escape routes should be prominently marked. A single step on the line of a doorway is acceptable, subject to paragraph 3.107.

Fixed ladders

3.85 Fixed ladders should not be provided as a means of escape for members of the public. They should only be provided where a conventional stair is impractical, such as for access to plant rooms which are not normally occupied.

Helical stairs and spiral stairs

3.86 Helical stairs and spiral stairs may form part of an escape route provided they are designed in accordance with **BS 5395-2**. If they are intended to serve members of the public, stairs should be type E (public) stairs.

Fire resistance of doors

3.87 Fire resistance test criteria are set out in Appendix C. Standards of performance are summarised in Table C1.

Fire resistance of glazed elements

3.88 If glazed elements in fire resisting enclosures and doors can only meet the required integrity performance, their use is limited. These limitations depend on whether the enclosure forms part of a protected shaft (see Section 7) and the provisions set out in Appendix B, Table B5. If both integrity and insulation performance can be met, there is no restriction in this document on the use or amount of glass.

3.89 Glazed elements should also comply with the following, where necessary.

a. If the enclosure forms part of a protected shaft: Section 7.

b. Appendix B, Table B5.

c. Guidance on the safety of glazing: Approved Document K.

Doors on escape routes

3.90 Doors should be readily openable to avoid undue delay to people escaping. Doors on escape routes (both within and from the building) should comply with paragraphs 3.91 to 3.98. Guidance on door closing and 'hold open' devices for fire doorsets is set out in Appendix C.

NOTE: Paragraphs 3.91 to 3.98 do not apply to flat entrance doors.

Door fastenings

3.91 In general, doors on escape routes (whether or not the doors are fire doorsets) should be either of the following.

a. Not fitted with a lock, latch or bolt fastenings.

b. Fitted only with simple fastenings that are all of the following.

i.   Easy to operate; it should be apparent how to undo the fastening.

ii.   Operable from the side approached by people escaping.

iii.   Operable without a key.

iv.   Operable without requiring people to manipulate more than one mechanism.

Doors may be fitted with hardware to allow them to be locked when rooms are empty.

If a secure door is operated by code or combination keypad, swipe or proximity card, biometric data, etc., a security mechanism override should be possible from the side approached by people escaping.

3.92 Electrically powered locks should return to the unlocked position in all of the following situations.

a.   If the fire detection and alarm system operates.

b.   If there is loss of power or system error.

c.   If the security mechanism override is activated.

Security mechanism overrides for electrically powered locks should be a Type A call point, as described in **BS 7273-4**. The call point should be positioned on the side approached by people escaping. If the door provides escape in either direction, a call point should be installed on both sides of the door.

3.93 Guidance on door closing and 'hold open' devices for fire doorsets is set out in Appendix C.

Direction of opening

3.94 The door of any doorway or exit should be hung to open in the direction of escape whenever reasonably practicable. It should always be hung to open in the direction of escape if more than 60 people might be expected to use it during a fire.

Amount of opening and effect on associated escape routes

3.95 All doors on escape routes should be hung to meet both of the following conditions.

a.   Open by a minimum of 90 degrees.

b.   Open with a swing that complies with both of the following.

i.   Is clear of any change of floor level, other than a threshold or single step on the line of the doorway.

ii.   Does not reduce the effective width of any escape route across a landing.

3.96 Any door opening towards a corridor or a stair should be recessed to prevent its swing encroaching on the effective width.

Vision panels in doors

3.97 Doors should contain vision panels in both of the following situations.

a.   Where doors on escape routes divide corridors.

b.   Where doors are hung to swing both ways.

Approved Document M contains guidance about vision panels in doors across accessible corridors and Approved Document K contains guidance about the safety of glazing.

Revolving and automatic doors

3.98 Where revolving doors, automatic doors and turnstiles are placed across escape routes they should comply with one of the following.

    a.   They are automatic doors of the required width and comply with one of the following conditions.

        i.   Their failsafe system provides outward opening from any open position.

        ii.   They have a monitored failsafe system to open the doors if the mains electricity supply fails.

        iii.   They failsafe to the open position if the power fails.

    b.   Non-automatic swing doors of the required width are provided immediately adjacent to the revolving or automatic door or turnstile.

Lifts

Fire protection of lift installations

3.99 Lift wells should comply with one of the following conditions.

    a.   Be sited within the enclosures of a protected stairway.

    b.   Be enclosed with fire resisting construction (minimum REI 30) when in a position that might prejudice the means of escape.

3.100 A lift well connecting different compartments should form a protected shaft (see Section 7).

3.101 In buildings designed for phased evacuation or progressive horizontal evacuation, if the lift well is not within the enclosures of a protected stairway, its entrance should be separated at every storey by a protected lobby (minimum REI 30).

3.102 In basements and enclosed car parks, the lift should be within the enclosure of a protected stairway. Otherwise, the lift should be approached only via a protected lobby or protected corridor (minimum REI 30).

3.103 If a lift delivers into a protected corridor or protected lobby serving sleeping accommodation and also serves a storey containing a high fire risk (such as a kitchen, communal areas, stores, etc.) then the lift should be separated from the high fire risk area(s) by a protected lobby or protected corridor (minimum REI 30).

3.104 A lift shaft serving storeys above ground level should not serve any basement, if either of the following applies.

    a.   There is only one escape stair serving storeys above ground level and smoke from a basement fire would adversely affect escape routes in the upper storeys.

    b.   The lift shaft is within the enclosure to an escape stair that terminates at ground level.

3.105 Lift machine rooms should be sited over the lift well where possible. Where buildings or part of a building with only one stairway make this arrangement impractical, the lift machine room should be sited outside the protected stairway.

Final exits

3.106 People should be able to rapidly leave the area around the building. Direct access to a street, passageway, walkway or open space should be available. The route away from the building should comply with the following.

a. Be well defined.

b. If necessary, have suitable guarding.

3.107 Final exits should not present a barrier for disabled people. Where the route to a final exit does not include stairs, a level threshold and, where necessary, a ramp should be provided.

3.108 Final exit locations should be clearly visible and recognisable.

3.109 Final exits should avoid outlets of basement smoke vents and openings to transformer chambers, refuse chambers, boiler rooms and similar risks.

B2

Requirement B2: Internal fire spread (linings)

This section deals with the following requirement from Part B of Schedule 1 to the Building Regulations 2010.

Requirement

Requirement	Limits on application
Internal fire spread (linings)	
B2. (1) To inhibit the spread of fire within the building, the internal linings shall—	
(a) adequately resist the spread of flame over their surfaces; and	
(b) have, if ignited, either a rate of heat release or a rate of fire growth, which is reasonable in the circumstances.	
(2) In this paragraph "internal linings" means the materials or products used in lining any partition, wall, ceiling or other internal structure.	

Intention

In the Secretary of State's view, requirement B2 is met by achieving a restricted spread of flame over internal linings. The building fabric should make a limited contribution to fire growth, including a low rate of heat release.

It is particularly important in circulation spaces, where linings may offer the main means by which fire spreads and where rapid spread is most likely to prevent occupants from escaping.

Requirement B2 *does not* include guidance on the following.

a. Generation of smoke and fumes.

b. The upper surfaces of floors and stairs.

c. Furniture and fittings.

Section 4: Wall and ceiling linings

Classification of linings

4.1 The surface linings of walls and ceilings should meet the classifications in Table 4.1.

Table 4.1 Classification of linings

Location	Classification
Small rooms of maximum internal floor area of 4m^2	D-s3, d2
Garages (as part of a dwellinghouse) of maximum internal floor area of 40m^2	
Other rooms (including garages)	C-s3, d2
Circulation spaces within a dwelling	
Other circulation spaces (including the common areas of blocks of flats)	B-s3, d2[1]

NOTE:

1. Wallcoverings which conform to **BS EN 15102**, achieving at least class C-s3, d2 and bonded to a class A2-s3, d2 substrate, will also be acceptable.

Walls

4.2 For the purposes of this requirement, a wall includes both of the following.

    a. The internal surface of internal and external glazing (except glazing in doors).

    b. Any part of a ceiling which slopes at an angle greater than 70 degrees to the horizontal.

4.3 For the purposes of this requirement, a wall *does not* include any of the following.

    a. Doors and door frames.

    b. Window frames and frames in which glazing is fitted.

    c. Architraves, cover moulds, picture rails, skirtings and similar narrow members.

    d. Fireplace surrounds, mantle shelves and fitted furniture.

4.4 Parts of walls in rooms may be of lower performance than stated in Table 4.1, but no worse than class D-s3, d2. In any one room, the total area of lower performance wall lining should be less than an area equivalent to half of the room's floor area, up to a maximum of 20m^2 of wall lining.

Ceilings

4.5 For the purposes of this requirement, a ceiling includes all of the following.

    a. Glazed surfaces.

    b. Any part of a wall at 70 degrees or less to the horizontal.

    c. The underside of a gallery.

    d. The underside of a roof exposed to the room below.

4.6 For the purposes of this requirement, a ceiling *does not* include any of the following.

a. Trap doors and their frames.

b. The frames of windows or rooflights and frames in which glazing is fitted.

c. Architraves, cover moulds, picture rails, exposed beams and similar narrow members.

Rooflights

4.7 Rooflights should meet the following classifications, according to material. No guidance for European fire test performance is currently available, because there is no generally accepted test and classification procedure.

a. Non-plastic rooflights should meet the relevant classification in Table 4.1.

b. Plastic rooflights, if the limitations in Table 4.2 and Table 12.2 are observed, should be a minimum class D-s3, d2 rating. Otherwise they should meet the relevant classification in Table 4.1.

Special applications

4.8 Any flexible membrane covering a structure, other than an air-supported structure, should comply with Appendix A of **BS 7157**.

4.9 Guidance on the use of PTFE-based materials for tension-membrane roofs and structures is given in the BRE report BR 274.

Fire behaviour of insulating core panels used internally

4.10 Insulating core panels consist of an inner core of insulation sandwiched between, and bonded to, a membrane, such as galvanised steel or aluminium.

Where they are used internally they can present particular problems with regard to fire spread and should meet all of the following conditions.

a. Panels should be sealed to prevent exposure of the core to a fire. This includes at joints and where services penetrate the panel.

b. In high fire risk areas, such as kitchens, places of special fire hazard, or in proximity to where hot works occur, only class A1 cored panels should be used.

c. Fixing systems for all panels should be designed to take account of the potential for the panel to delaminate. For instance, where panels are used to form a suspended ceiling, the fixing should pass through the panel and support it from the lower face.

Other controls on internal surface properties

4.11 Guidance on the control of flame spread is given in the following sections.

a. Stairs and landings: Sections 2 and 3 (escape stairs) and Section 15 (firefighting shafts).

b. Exposed surfaces above fire-protecting suspended ceilings: Section 8.

c. Enclosures to above-ground drainage system pipes: Section 9.

Thermoplastic materials

General provisions

4.12 Thermoplastic materials that do not meet the classifications in Table 4.1 can be used as described in paragraphs 4.13 to 4.17. No guidance for European fire test performance is currently available, because there is no generally accepted test and classification procedure.

Thermoplastic materials are defined in Appendix B, paragraph B11. Classifications used here are explained in paragraph B13.

Windows

4.13 Thermoplastic material classified as a TP(a) rigid product may be used to glaze external windows to rooms, *but not* external windows to circulation spaces. Approved Document K includes guidance on the safety of glazing.

Rooflights

4.14 In rooms and circulation spaces other than protected stairways, rooflights may be constructed of thermoplastic material if they comply with both of the following.

a. The lower surface is classified as TP(a) rigid or TP(b).

b. The size and location of the rooflights follow the limits in Table 4.2, Table 12.2 and Table 12.3.

Lighting diffusers

4.15 The following paragraphs apply to lighting diffusers forming part of a ceiling. Diffusers may be part of a luminaire or used below sources of light. The following paragraphs *do not* apply to diffusers of light fittings attached to the soffit of a ceiling or suspended beneath a ceiling (Diagram 4.1).

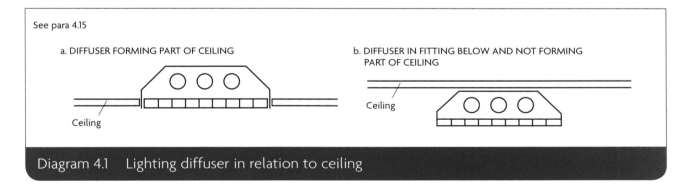

See para 4.15

a. DIFFUSER FORMING PART OF CEILING

b. DIFFUSER IN FITTING BELOW AND NOT FORMING PART OF CEILING

Ceiling

Ceiling

Diagram 4.1   Lighting diffuser in relation to ceiling

4.16 Diffusers constructed of thermoplastic material may be incorporated in ceilings to rooms and circulation spaces, but not to protected stairways, if both the following conditions are met.

a. Except for the upper surfaces of the thermoplastic panels, wall and ceiling surfaces exposed in the space above the suspended ceiling should comply with paragraph 4.1.

b. Diffusers should be classified as one of the following.

    i.   TP(a) rigid – no restrictions on their extent.

    ii.   TP(b) – limited in their extent (see Table 4.2 and Diagram 4.2).

Suspended or stretched-skin ceilings

4.17 A ceiling constructed from TP(a) flexible panels should meet the following conditions.

a. Have a maximum area of 5m².

b. Be supported on all sides.

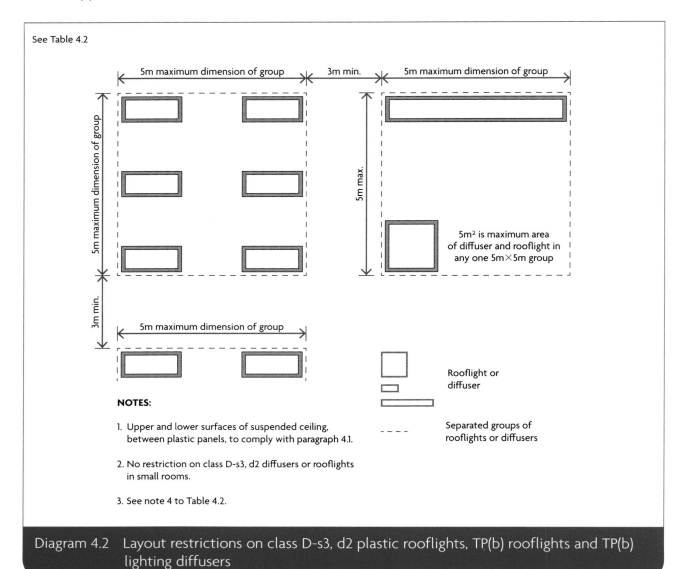

See Table 4.2

NOTES:

1. Upper and lower surfaces of suspended ceiling, between plastic panels, to comply with paragraph 4.1.

2. No restriction on class D-s3, d2 diffusers or rooflights in small rooms.

3. See note 4 to Table 4.2.

Diagram 4.2   Layout restrictions on class D-s3, d2 plastic rooflights, TP(b) rooflights and TP(b) lighting diffusers

Table 4.2 Limitations applied to thermoplastic rooflights and lighting diffusers in suspended ceilings and class D-s3, d2 plastic rooflights[1]

Minimum classification of lower surface	Use of space below the diffusers or rooflights	Maximum area of each diffuser or rooflight[2] (m²)	Maximum total area of diffusers and rooflights as a percentage of floor area of the space in which the ceiling is located (%)	Minimum separation distance between diffusers or rooflights[2] (m)
TP(a)	Any except protected stairways	No limit[3]	No limit	No limit
Class D-s3, d2[4] or TP(b)	Rooms	5	50[5]	3
	Circulation spaces except protected stairways	5	15[5]	3

NOTES:

1. This table does not apply to products that meet the provisions in Table 4.1.

2. Smaller rooflights and diffusers can be grouped together provided that both of the following satisfy the dimensions in Diagram 4.2 or 4.3.

   a. The overall size of the group.

   b. The space between one group and any others.

3. Lighting diffusers of TP(a) flexible rating should be used only in panels of a maximum of 5m² each. See paragraph 4.17.

4. There are no limits on the use of class D-s3, d2 materials in small rooms. See Table 4.1.

5. The minimum 3m separation given in Diagram 4.2 between each 5m² group must be maintained. Therefore, in some cases, it may not be possible to use the maximum percentage quoted.

See Table 4.2

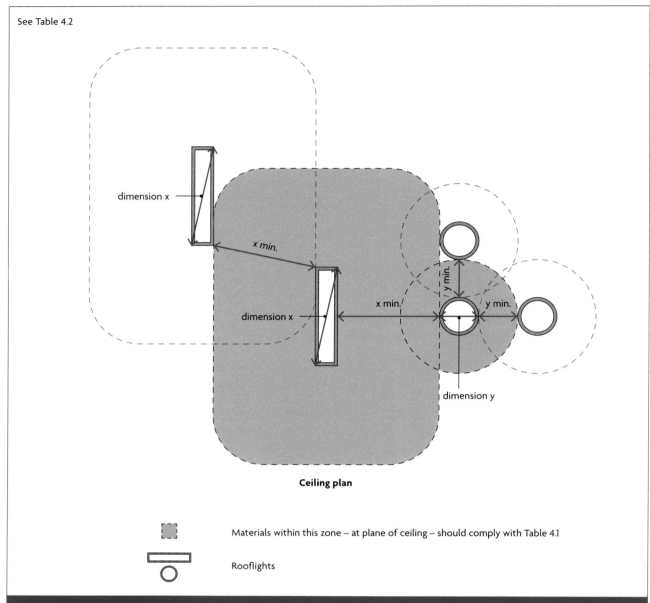

Ceiling plan

Materials within this zone – at plane of ceiling – should comply with Table 4.1

Rooflights

Diagram 4.3 Layout restrictions on small class D-s3, d2 plastic rooflights, TP(b) rooflights and lighting diffusers

Requirement B3: Internal fire spread (structure)

These sections deal with the following requirement from Part B of Schedule 1 to the Building Regulations 2010.

Requirement

Requirement	Limits on application
Internal fire spread (structure)	
B3. (1) The building shall be designed and constructed so that, in the event of fire, its stability will be maintained for a reasonable period	
(2) A wall common to two or more buildings shall be designed and constructed so that it adequately resists the spread of fire between those buildings. For the purposes of this sub-paragraph a house in a terrace and a semi-detached house are each to be treated as a separate building.	
(3) Where reasonably necessary to inhibit the spread of fire within the building, measures shall be taken, to an extent appropriate to the size and intended use of the building, comprising either or both of the following—	Requirement B3(3) does not apply to material alterations to any prison provided under section 33 of the Prison Act 1952.
(a) sub-division of the building with fire-resisting construction;	
(b) installation of suitable automatic fire suppression systems.	
(4) The building shall be designed and constructed so that the unseen spread of fire and smoke within concealed spaces in its structure and fabric is inhibited.	

Intention

In the Secretary of State's view, requirement B3 is met by achieving all of the following.

a. For defined periods, loadbearing elements of structure withstand the effects of fire without loss of stability.

b. Compartmentation of buildings by fire resisting construction elements.

c. Automatic fire suppression is provided where it is necessary.

d. Protection of openings in fire-separating elements to maintain continuity of the fire separation.

e. Inhibition of the unseen spread of fire and smoke in cavities, in order to reduce the risk of structural failure and spread of fire and smoke, where they pose a threat to the safety of people in and around the building.

The extent to which any of these measures are necessary is dependent on the use of the building and, in some cases, its size, and on the location of the elements of construction.

Section 5: Internal fire spread – dwellinghouses

Loadbearing elements of structure

Fire resistance standard

5.1 Elements such as structural frames, beams, columns, loadbearing walls (internal and external), floor structures and gallery structures should have, as a minimum, the fire resistance given in Appendix B, Table B3.

5.2 If one element of structure supports or stabilises another, as a minimum the supporting element should have the same fire resistance as the other element.

5.3 The following are excluded from the definition of 'element of structure'.

    a. A structure that supports only a roof, unless either of the following applies.

        i.   The roof performs the function of a floor, such as a roof terrace, or as a means of escape.

        ii. The structure is essential for the stability of an external wall that needs to be fire resisting (e.g. to achieve compartmentation or for the purposes of preventing fire spread between buildings).

    b. The lowest floor of the building.

    c. External walls, such as curtain walls or other forms of cladding, which transmit only self weight and wind loads and do not transmit floor load.

NOTE: In some cases, structural members within a roof may be essential for the structural stability system of the building. In these cases, the structural members in the roof do not just support a roof and must demonstrate the relevant fire resistance for the building as required by paragraph 5.2 above.

Floors in loft conversions

5.4 Where adding an additional storey to a two storey single family dwellinghouse, new floors should have a minimum REI 30 fire resistance. Any floor forming part of the enclosure to the circulation space between the loft conversion and the final exit should achieve a minimum rating of REI 30.

The existing first-storey construction should have a minimum rating of R 30. The fire performance may be reduced for integrity and insulation, when both of the following conditions are met.

    a. Only one storey is added, containing a maximum of two habitable rooms.

    b. The new storey has a maximum total area of 50m².

Compartmentation

Provision of compartmentation

5.5 Dwellinghouses that are semi-detached or in terraces should be considered as separate buildings. Every wall separating the dwellinghouses should be constructed as a compartment wall (see paragraphs 5.8 to 5.12).

5.6 If a garage is attached to or forms an integral part of a dwellinghouse, the garage should be separated from the rest of the dwellinghouse by fire resisting construction (minimum REI 30) (Diagram 5.1).

5.7 Where a door is provided between a dwellinghouse and the garage (see Diagram 5.1), it should meet one of the following conditions.

   a. The garage floor should be laid such that it falls away from the door to the outside, to allow fuel spills to flow away.

   b. The door opening should be a minimum of 100mm above the level of the garage floor.

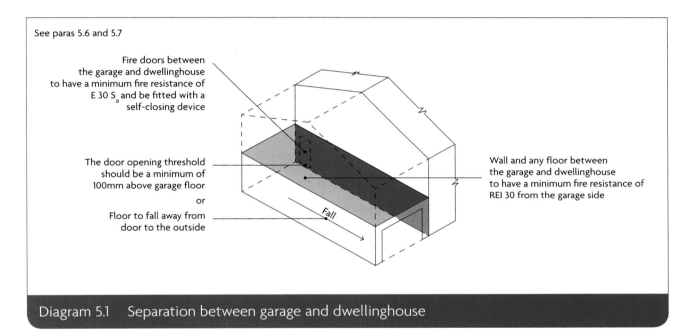

See paras 5.6 and 5.7

Fire doors between the garage and dwellinghouse to have a minimum fire resistance of E 30 S$_a$ and be fitted with a self-closing device

The door opening threshold should be a minimum of 100mm above garage floor

or

Floor to fall away from door to the outside

Fall

Wall and any floor between the garage and dwellinghouse to have a minimum fire resistance of REI 30 from the garage side

Diagram 5.1   Separation between garage and dwellinghouse

Construction of compartment walls and compartment floors

General provisions

5.8 All compartment walls and compartment floors should achieve both of the following.

   a. Form a complete barrier to fire between the compartments they separate.

   b. Have the appropriate fire resistance, as given in Appendix B, Table B3 and Table B4.

5.9 Timber beams, joists, purlins and rafters may be built into or carried through a masonry or concrete compartment wall if the openings for them are both of the following.

   a. As small as practicable.

   b. Fire-stopped.

If trussed rafters bridge the wall, failure of the truss due to a fire in one compartment should not cause failure of the truss in another compartment.

Compartment walls between buildings

5.10 Adjoining buildings should only be separated by walls, not floors. Compartment walls common to two or more buildings should comply with both of the following.

    a. Run the full height of the building in a continuous vertical plane.

    b. Be continued through any roof space to the underside of the roof (see Diagram 5.2).

Junction of compartment wall with roof

5.11 A compartment wall should achieve both of the following.

    a. Meet the underside of the roof covering or deck, with fire-stopping to maintain the continuity of fire resistance.

    b. Be continued across any eaves.

5.12 To reduce the risk of fire spreading over the roof from one compartment to another, a 1500mm wide zone of the roof, either side of the wall, should have a covering classified as $B_{ROOF}(t4)$, on a substrate or deck of a material rated class A2-s3, d2 or better, as set out in Diagram 5.2a.

Thermoplastic rooflights that, because of paragraph 12.7, are regarded as having a $B_{ROOF}(t4)$ classification are *not* suitable for use in that zone.

5.13 Materials achieving class B-s3, d2 or worse used as a substrate to the roof covering and any timber tiling battens, fully bedded in mortar or other suitable material for the width of the wall (Diagram 5.2b), may extend over the compartment wall in buildings that are a maximum of 15m high.

5.14 Double-skinned insulated roof sheeting should incorporate a band of material rated class A2-s3, d2 or better, a minimum of 300mm in width, centred over the wall.

5.15 As an alternative to the provisions of paragraphs 5.12 to 5.14, the compartment wall may extend through the roof for a minimum of either of the following (see Diagram 5.2c).

    a. Where the height difference between the two roofs is less than 375mm, 375mm above the top surface of the adjoining roof covering.

    b. 200mm above the top surface of the adjoining roof covering where either of the following applies.

        i.   The height difference between the two roofs is 375mm or more.

        ii.   The roof coverings either side of the wall are of a material classified as $B_{ROOF}(t4)$.

See paras 5.12 to 5.15

a. ANY BUILDING OR COMPARTMENT

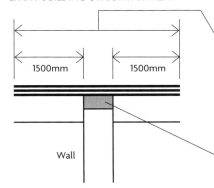

1500mm	1500mm

Wall

Roof covering over this distance to be designated B_{ROOF}(t4) rated on deck of material of class A2-s3, d2 or better. Roof covering and deck could be composite structure, e.g. profiled steel cladding.

Double-skinned insulated roof sheeting should incorporate a band of material rated class A2-s3, d2 or better, a minimum of 300mm in width, centred over the wall.

If roof support members pass through the wall, fire protection to these members for a distance of 1500mm on either side of the wall may be needed to delay distortion at the junction (see paragraph 5.9).

Fire-stopping to be carried up to underside of roof covering, e.g. roof tiles.

b. RESIDENTIAL (DWELLINGS) AND RESIDENTIAL (OTHER) A MAXIMUM OF 15M HIGH

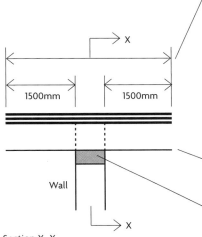

1500mm	1500mm

Wall

Section X–X

Roof covering to be designated B_{ROOF}(t4) rated for at least this distance.

Boarding (used as a substrate) or timber tiling battens may be carried over the wall provided that they are fully bedded in mortar (or other no less suitable material) where over the wall.

Thermoplastic insulation materials should not be carried over the wall.

Double-skinned insulated roof sheeting with a thermoplastic core should incorporate a band of material of class A2-s3, d2 at least 300mm wide centred over the wall.

Sarking felt may also be carried over the wall.

If roof support members pass through the wall, fire protection to these members for a distance of 1500mm on either side of the wall may be needed to delay distortion at the junction (see paragraph 5.9).

Fire-stopping to be carried up to underside of roof covering, boarding or slab.

Roof covering to be designated B_{ROOF}(t4) rated for at least 1500mm either side of wall.

Roofing battens and sarking felt may be carried over the wall.

Fire-stopping to be carried up to underside of roof covering above and below sarking felt.

NOTES:
1. Fire-stopping should be carried over the full thickness of the wall.
2. Fire-stopping should be extended into any eaves.
3. The compartment wall does not necessarily need to be constructed of masonry.

c. ANY BUILDING OR COMPARTMENT

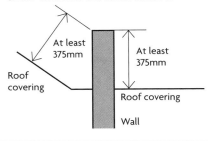

At least 375mm

At least 375mm

Roof covering

Roof covering

Wall

The wall should be extended up through the roof for a height of at least 375mm above the top surface of the adjoining roof covering.

Where there is a height difference of at least 375 mm between two roofs or where the roof coverings on either side of the wall are B_{ROOF}(t4) rated, the height of the upstand/parapet wall above the highest roof may be reduced to 200mm.

Diagram 5.2   Junction of compartment wall with roof

B3

Cavities

5.16 Cavities in the construction of a building provide a ready route for the spread of smoke and flame, which can present a greater danger as any spread is concealed. For the purpose of this document, a cavity is considered to be any concealed space.

Provision of cavity barriers

5.17 To reduce the potential for fire spread, cavity barriers should be provided for both of the following.

a. To divide cavities.

b. To close the edges of cavities.

Cavity barriers should not be confused with fire-stopping details (Section 9).

5.18 Cavity barriers should be provided at all of the following locations.

a. At the edges of cavities, including around openings (such as windows, doors and exit/entry points for services).

b. At the junction between an external cavity wall and every compartment floor and compartment wall.

c. At the junction between an internal cavity wall and every compartment floor, compartment wall or other wall or door assembly forming a fire resisting barrier.

This does not apply where a wall meets the conditions of Diagram 5.3.

5.19 It is not appropriate to complete a line of compartment walls by fitting cavity barriers above them. The compartment wall should be extended to the underside of the floor or roof above.

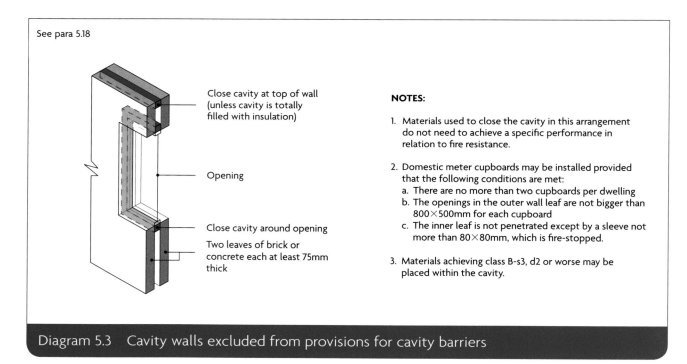

See para 5.18

Close cavity at top of wall (unless cavity is totally filled with insulation)

Opening

Close cavity around opening

Two leaves of brick or concrete each at least 75mm thick

NOTES:

1. Materials used to close the cavity in this arrangement do not need to achieve a specific performance in relation to fire resistance.

2. Domestic meter cupboards may be installed provided that the following conditions are met:
   a. There are no more than two cupboards per dwelling
   b. The openings in the outer wall leaf are not bigger than 800×500mm for each cupboard
   c. The inner leaf is not penetrated except by a sleeve not more than 80×80mm, which is fire-stopped.

3. Materials achieving class B-s3, d2 or worse may be placed within the cavity.

Diagram 5.3   Cavity walls excluded from provisions for cavity barriers

Construction and fixings for cavity barriers

5.20 Cavity barriers, tested from each side separately, should provide a minimum of both of the following:

a. 30 minutes' integrity (E 30)

b. 15 minutes' insulation (I 15).

They may be formed by a construction provided for another purpose if it achieves the same performance.

5.21 Cavity barriers in a stud wall or partition, or provided around openings, may be formed of any of the following.

a. Steel, a minimum of 0.5mm thick.

b. Timber, a minimum of 38mm thick.

c. Polythene-sleeved mineral wool, or mineral wool slab, under compression when installed in the cavity.

d. Calcium silicate, cement-based or gypsum-based boards, a minimum of 12mm thick.

These do not necessarily achieve the performance specified in paragraph 5.20.

NOTE: Cavity barriers provided around openings may be formed by the window or door frame, if the frame is constructed of steel or timber of the minimum thickness in (a) or (b), as appropriate.

5.22 Cavity barriers should be tightly fitted to a rigid construction and mechanically fixed in position. If this is not possible (e.g. where a cavity barrier joins to slates, tiles, corrugated sheeting or similar materials) the junction should be fire-stopped.

5.23 Cavity barriers should be fixed so their performance is unlikely to be made ineffective by any of the following.

a. Movement of the building due to subsidence, shrinkage or temperature change, and movement of the external envelope due to wind.

b. During a fire, collapse of services penetrating the cavity barriers, either by the failure of the supporting system or through degradation of the service itself (e.g. by melting or burning).

c. During a fire, failure of the cavity barrier fixings. (In roof spaces, where cavity barriers are fixed to roof members, there is no expectation of fire resistance from roof members provided for the purpose of support.)

d. During a fire, failure of any material or construction to which cavity barriers abut. (For example, a suspended ceiling that continues over a fire resisting wall or partition collapses, and the cavity barrier fails prematurely because the ceiling was not designed to provide a minimum fire resistance of EI 30.)

Openings in cavity barriers

5.24 Openings should be limited to the following.

a. Fire doorsets with a minimum E 30 rating, fitted in accordance with Appendix C.

b. The passage of pipes that follow the provisions in Section 9.

c. The passage of cables or conduits containing one or more cables.

d. Openings fitted with a suitably mounted and appropriate fire damper.

e. Ducts that are either of the following.

    i.   Fire resisting (minimum E 30).

    ii.   Fitted with a suitably mounted and appropriate fire damper where they pass through the cavity barrier.

NOTE: For further guidance on openings in cavity barriers see Section 9.

Section 6: Loadbearing elements of structures – flats

Fire resistance standard

6.1 Elements such as structural frames, beams, columns, loadbearing walls (internal and external), floor structures and gallery structures should have, as a minimum, the fire resistance given in Appendix B, Table B3.

NOTE: If one element of structure supports or stabilises another, as a minimum the supporting element should have the same fire resistance as the other element.

6.2 The following are excluded from the definition of 'element of structure'.

a. A structure that supports only a roof, unless either of the following applies.

  i. The roof performs the function of a floor, such as for parking vehicles, or as a means of escape.

  ii. The structure is essential for the stability of an external wall that needs to be fire resisting (e.g. to achieve compartmentation or for the purposes of preventing fire spread between buildings).

b. The lowest floor of the building.

c. A platform floor.

d. External walls, such as curtain walls or other forms of cladding, which transmit only self weight and wind loads and do not transmit floor load.

NOTE: In some cases, structural members within a roof may be essential for the structural stability system of the building. In these cases, the structural members in the roof do not just support a roof and must demonstrate the relevant fire resistance for the building as required by the note to paragraph 6.1 above.

Additional guidance

6.3 If a loadbearing wall is any of the following, guidance in other sections may also apply.

a. A compartment wall (including a wall common to two buildings): Section 7.

b. Enclosing a place of special fire hazard: Section 7.

c. Protecting a means of escape: Sections 2 and 3.

d. An external wall: Sections 10 and 11.

e. Enclosing a firefighting shaft: Section 15.

6.4 If a floor is also a compartment floor, see Section 7.

Conversion to flats

6.5 Where an existing dwellinghouse or other building is converted into flats, a review of the existing construction should be carried out. Retained timber floors may make it difficult to meet the relevant provisions for fire resistance.

6.6 In a converted building with a maximum of three storeys, a minimum REI 30 fire resistance could be accepted for elements of structure if the means of escape conform to the provisions of Section 3.

6.7 In a converted building with four or more storeys, the full standard of fire resistance given in Appendix B is necessary.

Section 7: Compartmentation/sprinklers — flats

Provision of compartmentation

7.1 All of the following should be provided as compartment walls and compartment floors and should have, as a minimum, the fire resistance given in Appendix B, Table B3.

a. Any floor and wall separating a flat from another part of the building.

b. Any wall enclosing a refuse storage chamber.

c. Any wall common to two or more buildings.

Places of special fire hazard

7.2 Fire resisting construction enclosing these places should achieve minimum REI 30. These walls and floors are not compartment walls and compartment floors.

7.3 Parts of a building occupied mainly for different purposes should be separated from one another by compartment walls and/or compartment floors. Compartmentation is not needed if one of the different purposes is ancillary to the other. See paragraphs 0.18 and 0.19.

Sprinklers

7.4 Blocks of flats with a floor more than 30m above ground level should be fitted with a sprinkler system throughout the building in accordance with Appendix E.

NOTE: Sprinklers are not required in the common areas such as stairs, corridors or landings when these areas are fire sterile.

Construction of compartment walls and compartment floors

General provisions

7.5 All compartment walls and compartment floors should achieve both of the following.

a. Form a complete barrier to fire between the compartments they separate.

b. Have the appropriate fire resistance, as given in Appendix B, Tables B3 and B4.

7.6 Timber beams, joists, purlins and rafters may be built into or carried through a masonry or concrete compartment wall if the openings for them are both of the following.

a. As small as practicable.

b. Fire-stopped.

If trussed rafters bridge the wall, failure of the truss due to a fire in one compartment should not cause failure of the truss in another compartment.

7.7 Where services could provide a source of ignition, the risk of fire developing and spreading into adjacent compartments should be controlled.

Compartment walls between buildings

7.8 Adjoining buildings should only be separated by walls, not floors. Compartment walls common to two or more buildings should comply with both of the following.

    a. Run the full height of the building in a continuous vertical plane.

    b. Be continued through any roof space to the underside of the roof (see Diagram 5.2).

Separated parts of buildings

7.9 Compartment walls forming a separated part of a building should run the full height of the building in a continuous vertical plane.

Separated parts can be assessed independently to determine the appropriate standard of fire resistance in each. The two separated parts can have different standards of fire resistance.

Other compartment walls

7.10 Compartment walls not described in paragraphs 7.8 and 7.9 should run the full height of the storey in which they are situated.

7.11 Compartment walls in a top storey beneath a roof should be continued through the roof space.

Junction of compartment wall or compartment floor with other walls

7.12 At the junction with another compartment wall or an external wall, the fire resistance of the compartmentation should be maintained. Fire-stopping that meets the provisions in paragraphs 9.24 to 9.29 should be provided.

7.13 At the junction of a compartment floor and an external wall with no fire resistance, the external wall should be restrained at floor level. The restraint should reduce movement of the wall away from the floor if exposed to fire.

7.14 Compartment walls should be able to accommodate deflection of the floor, when exposed to fire, by either of the following means.

    a. Between the wall and floor, provide a head detail that is capable of maintaining its integrity while deforming.

    b. Design the wall so it maintains its integrity by resisting the additional vertical load from the floor above.

Where compartment walls are located within the middle half of a floor between vertical supports, the deflection may be assumed to be 40mm unless a smaller value can be justified by assessment. Outside this area, the limit can be reduced linearly to zero at the supports.

For steel beams that do not have the required fire resistance, reference should be made to SCI Publication P288.

Junction of compartment wall with roof

7.15 The requirements are the same as for dwellinghouses, detailed in paragraphs 5.11 and 5.12.

7.16 Materials achieving class B-s3, d2 or worse used as a substrate to the roof covering and any timber tiling battens, fully bedded in mortar or other suitable material for the width of the wall (Diagram 5.2b), may extend over the compartment wall in buildings that are both of the following.

    a. A maximum of 15m high.

b. In one of the following purpose groups.

    i. All residential purpose groups (purpose groups 1 and 2) other than 'residential (institutional)' (purpose group 2(a)).

    ii. 'Office' (purpose group 3).

    iii. 'Assembly and recreation' (purpose group 5).

7.17 Double-skinned insulated roof sheeting with a thermoplastic core should incorporate a band of material rated class A2-s3, d2 or better, a minimum of 300mm in width, centred over the wall.

7.18 As an alternative to the provisions of paragraph 7.16 or 7.17, the compartment wall may extend through the roof for a minimum of either of the following (see Diagram 5.2c).

a. Where the height difference between the two roofs is less than 375mm, 375mm above the top surface of the adjoining roof covering.

b. 200mm above the top surface of the adjoining roof covering where either of the following applies.

    i. The height difference between the two roofs is 375mm or more.

    ii. The roof coverings either side of the wall are of a material classified as $B_{ROOF}(t4)$.

Openings in compartmentation

Openings in compartment walls separating buildings or occupancies

7.19 Openings in a compartment wall common to two or more buildings should be limited to those for either of the following.

a. A fire doorset providing a means of escape, which has the same fire resistance as the wall and is fitted in accordance with the provisions in Appendix C.

b. The passage of a pipe that complies with the provisions in Section 9.

Openings in other compartment walls, or in compartment floors

7.20 Openings should be limited to those for any of the following.

a. Fire doorsets of the appropriate fire resistance, fitted in accordance with the provisions in Appendix C.

b. Pipes, ventilation ducts, service cables, chimneys, appliance ventilation ducts or ducts encasing one or more flue pipes, complying with the provisions in Section 9.

c. Refuse chutes of class A1 construction.

d. Atria designed in accordance with Annexes B and C of **BS 9999**.

e. Protected shafts that conform to the provisions in the following paragraphs.

Protected shafts

7.21 Stairs and service shafts connecting compartments should be protected to restrict the spread of fire between the compartments. These are called protected shafts. Walls or floors surrounding a protected shaft are considered to be compartment walls or compartment floors.

7.22 Any stair or other shaft passing directly from one compartment to another should be enclosed in a protected shaft. Protected shafts should be used for the following only, but may also include sanitary accommodation and washrooms.

    a. Stairs.

    b. Lifts.

    c. Escalators.

    d. Chutes.

    e. Ducts.

    f. Pipes.

    g. Additional provisions apply for both of the following.

        i. Protected shafts that are protected stairways: Sections 2 to 4.

        ii. Stairs that are also firefighting stairs: Section 15.

Construction of protected shafts

7.23 The construction enclosing a protected shaft (Diagram 7.1) should do all of the following.

    a. Form a complete barrier to fire between the compartments connected by the shaft.

    b. Have the appropriate fire resistance given in Appendix B, Table B3, *except for uninsulated glazed screens that meet the provisions of paragraph 7.24.*

    c. Satisfy the provisions for ventilation and the treatment of openings in paragraphs 7.28 and 7.29.

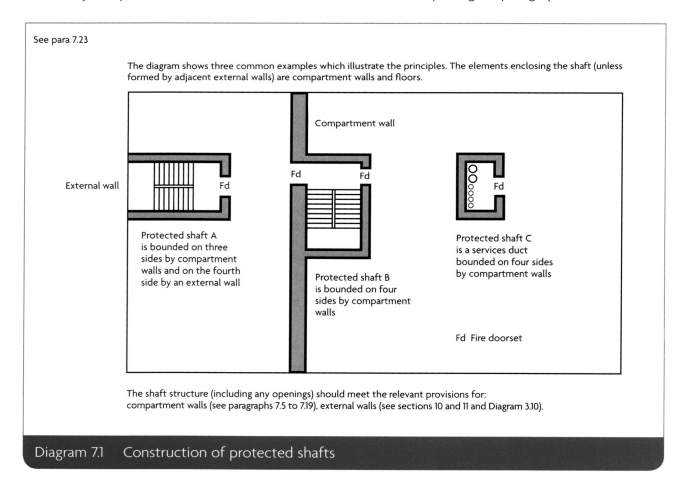

See para 7.23

The diagram shows three common examples which illustrate the principles. The elements enclosing the shaft (unless formed by adjacent external walls) are compartment walls and floors.

External wall

Compartment wall

Fd

Fd

Fd

Fd

Protected shaft A is bounded on three sides by compartment walls and on the fourth side by an external wall

Protected shaft B is bounded on four sides by compartment walls

Protected shaft C is a services duct bounded on four sides by compartment walls

Fd Fire doorset

The shaft structure (including any openings) should meet the relevant provisions for: compartment walls (see paragraphs 7.5 to 7.19), external walls (see sections 10 and 11 and Diagram 3.10).

Diagram 7.1   Construction of protected shafts

Uninsulated glazed screens to protected shafts

7.24 An uninsulated glazed screen may be incorporated in the enclosure to a protected shaft between a stair and a lobby or corridor entered from the stair. The enclosure must conform to Diagram 7.2 and meet all of the following conditions.

a. The standard of fire resistance required for the protected stairway is not more than REI 60.

b. The glazed screen complies with the following.

   i. It achieves a minimum rating of E 30.

   ii. It complies with the guidance on limits on areas of uninsulated glazing in Appendix B, Table B5.

c. The lobby or corridor is enclosed with fire resisting construction achieving a minimum rating of REI 30.

7.25 Where the measures in Diagram 7.2 are not provided, then both of the following apply.

a. The enclosing walls should comply with Appendix B, Table B3.

b. The doors should comply with Appendix B, Table B5.

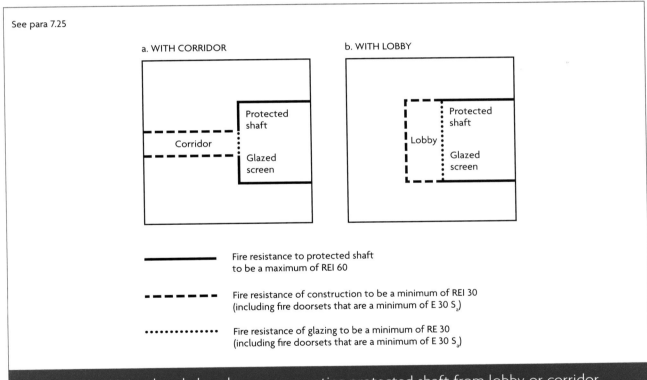

Diagram 7.2   Uninsulated glazed screen separating protected shaft from lobby or corridor

Pipes for oil or gas and ventilation ducts in protected shafts

7.26 A protected shaft containing a protected stairway and/or a lift *should not* also contain either of the following.

a. A pipe that conveys oil, other than in the mechanism of a hydraulic lift.

b. A ventilating duct. Two exceptions are as follows.

   i. A duct provided for pressurising the protected stairway to keep it smoke free.

ii.   A duct provided only to ventilate the protected stairway.

A pipe that is completely separated from a protected shaft by fire resisting construction is not considered to be contained within that shaft.

7.27   In a protected shaft, any pipe carrying natural gas or LPG should be both of the following.

a.   Of screwed steel or all-welded steel construction.

b.   Installed in accordance with both of the following.

i.   The Pipelines Safety Regulations 1996.

ii.   The Gas Safety (Installation and Use) Regulations 1998.

Ventilation of protected shafts conveying gas

7.28   A protected shaft conveying piped flammable gas should be ventilated direct to the outside air, by ventilation openings at high and low level in the shaft.

Any extension of the storey floor into the protected shaft should not compromise the free movement of air throughout the entire length of the shaft.

Guidance on shafts conveying piped flammable gas, including the size of ventilation openings, is given in **BS 8313**.

Openings into protected shafts

7.29   The external wall of a protected shaft does not normally need to have fire resistance. Situations where there are provisions are given in paragraph 3.63 (external walls of protected stairways, which may also be protected shafts) and paragraphs 15.8 to 15.11 (firefighting shafts).

Openings in other parts of the enclosure to a protected shaft should be limited to the following.

a.   If a wall common to two or more buildings forms part of the enclosure, only the following openings should be made in that wall.

i.   A fire doorset providing a means of escape, which has the same fire resistance as the wall and is fitted in accordance with the provisions in Appendix C.

ii.   The passage of a pipe that meets the provisions in Section 9.

b.   Other parts of the enclosure (other than an external wall) should only have openings for any of the following.

i.   Fire doorsets of the appropriate fire resistance, fitted in accordance with the provisions in Appendix C.

ii.   The passage of pipes which meet the provisions in Section 9.

iii.   Inlets to, outlets from and openings for a ventilation duct (if the shaft contains or serves as a ventilating duct), meeting the provisions in Section 9.

iv.   The passage of lift cables into a lift machine room (if the shaft contains a lift). If the machine room is at the bottom of the shaft, the openings should be as small as practicable.

Section 8: Cavities – flats

8.1 Cavities in the construction of a building provide a ready route for the spread of smoke and flame, which can present a greater danger as any spread is concealed. For the purpose of this document, a cavity is considered to be any concealed space.

Provision of cavity barriers

8.2 To reduce the potential for fire spread, cavity barriers should be provided for both of the following.

    a. To divide cavities.

    b. To close the edges of cavities.

See Diagram 8.1. Cavity barriers should not be confused with fire-stopping details (Section 9).

Pathways around fire-separating elements

Junctions and cavity closures

8.3 Cavity barriers should be provided at all of the following locations.

    a. At the edges of cavities, including around openings (such as windows, doors and exit/entry points for services).

    b. At the junction between an external cavity wall and every compartment floor and compartment wall.

    c. At the junction between an internal cavity wall and every compartment floor, compartment wall or other wall or door assembly forming a fire resisting barrier.

This does not apply where a wall meets the conditions of Diagram 8.2.

8.4 It is not appropriate to complete a line of compartment walls by fitting cavity barriers above them. The compartment walls should extend to the underside of the floor or roof above.

Protected escape routes

8.5 If the fire resisting construction of a protected escape route is either of the following.

    a. Not carried to full storey height.

    b. At the top storey, not carried to the underside of the roof covering.

Then the cavity *above or below* the fire resisting construction should be either of the following.

    i. Fitted with cavity barriers on the line of the enclosure.

    ii. For cavities above the fire resisting construction, enclosed on the lower side by a fire resisting ceiling (minimum EI 30) that extends throughout the building, compartment or separated part (see Diagram 8.3).

B3

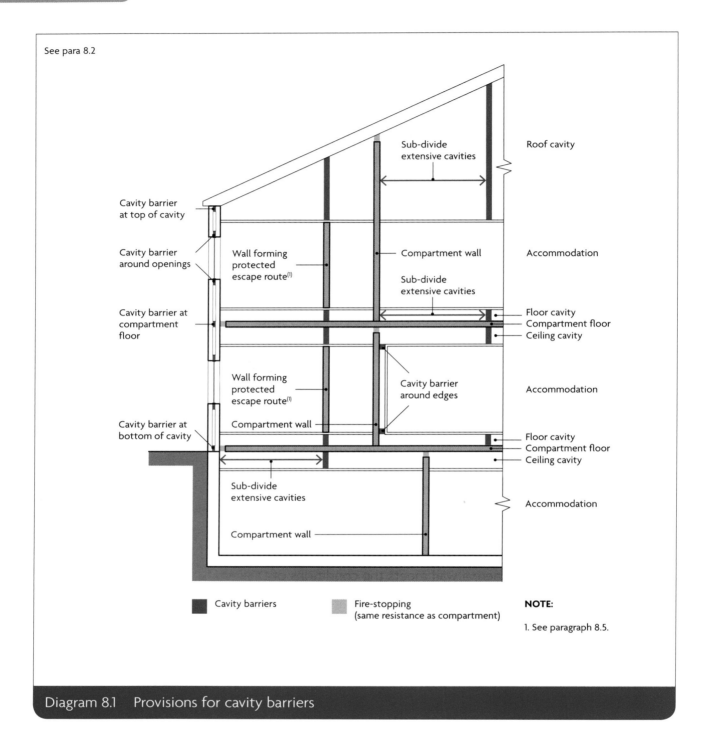

See para 8.2

Cavity barrier at top of cavity

Cavity barrier around openings

Wall forming protected escape route[1]

Cavity barrier at compartment floor

Wall forming protected escape route[1]

Cavity barrier at bottom of cavity

Compartment wall

Sub-divide extensive cavities

Sub-divide extensive cavities

Compartment wall

Sub-divide extensive cavities

Cavity barrier around edges

Compartment wall

Roof cavity

Accommodation

Floor cavity
Compartment floor
Ceiling cavity

Accommodation

Floor cavity
Compartment floor
Ceiling cavity

Accommodation

Cavity barriers

Fire-stopping (same resistance as compartment)

NOTE:

1. See paragraph 8.5.

Diagram 8.1   Provisions for cavity barriers

Cavities affecting alternative escape routes

8.6   In divided corridors (paragraph 3.25 and following) with cavities, fire-stopping should be provided to prevent alternative escape routes being affected by fire and/or smoke.

Double-skinned corrugated or profiled roof sheeting

8.7 Cavity barriers are not required between double-skinned corrugated or profiled insulated roof sheeting, if the sheeting complies with all of the following.

a. The sheeting is rated class A2-s3, d2 or better.

b. Both surfaces of the insulating layer are rated class C-s3, d2 or better.

c. Both surfaces of the insulating layer make contact with the inner and outer skins of cladding (Diagram 8.4).

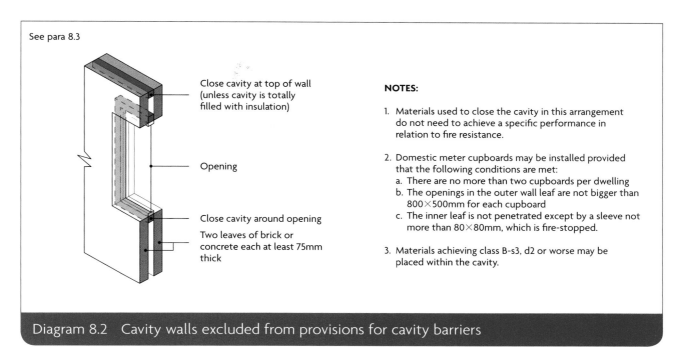

See para 8.3

Close cavity at top of wall (unless cavity is totally filled with insulation)

Opening

Close cavity around opening

Two leaves of brick or concrete each at least 75mm thick

NOTES:

1. Materials used to close the cavity in this arrangement do not need to achieve a specific performance in relation to fire resistance.

2. Domestic meter cupboards may be installed provided that the following conditions are met:
   a. There are no more than two cupboards per dwelling
   b. The openings in the outer wall leaf are not bigger than 800×500mm for each cupboard
   c. The inner leaf is not penetrated except by a sleeve not more than 80×80mm, which is fire-stopped.

3. Materials achieving class B-s3, d2 or worse may be placed within the cavity.

Diagram 8.2   Cavity walls excluded from provisions for cavity barriers

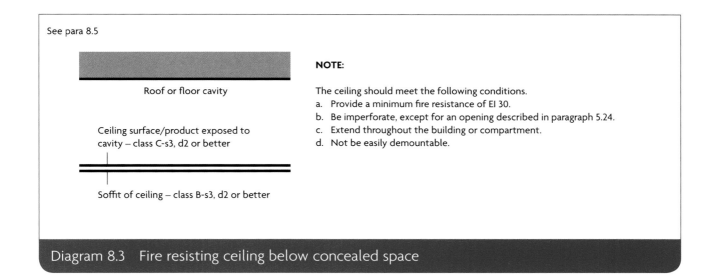

See para 8.5

Roof or floor cavity

Ceiling surface/product exposed to cavity – class C-s3, d2 or better

Soffit of ceiling – class B-s3, d2 or better

NOTE:

The ceiling should meet the following conditions.
a. Provide a minimum fire resistance of EI 30.
b. Be imperforate, except for an opening described in paragraph 5.24.
c. Extend throughout the building or compartment.
d. Not be easily demountable.

Diagram 8.3   Fire resisting ceiling below concealed space

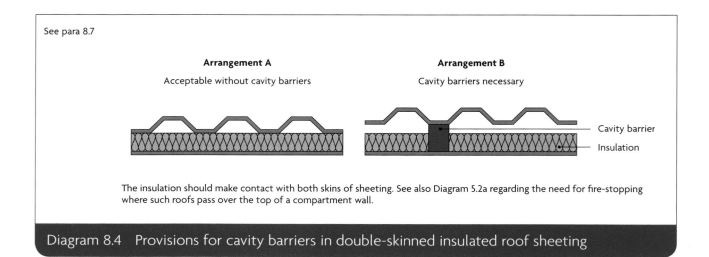

See para 8.7

Arrangement A

Acceptable without cavity barriers

Arrangement B

Cavity barriers necessary

Cavity barrier

Insulation

The insulation should make contact with both skins of sheeting. See also Diagram 5.2a regarding the need for fire-stopping where such roofs pass over the top of a compartment wall.

Diagram 8.4   Provisions for cavity barriers in double-skinned insulated roof sheeting

Construction and fixings for cavity barriers

8.8 Cavity barriers, tested from each side separately, should provide a minimum of both of the following:

a.   30 minutes' integrity (E 30)

b.   15 minutes' insulation (I 15).

They may be formed by a construction provided for another purpose if it achieves the same performance.

8.9 Cavity barriers should meet the requirements set out in paragraphs 5.21 to 5.23.

Section 9: Protection of openings and fire-stopping

Introduction

9.1 The performance of a fire-separating element should not be impaired. Every joint, imperfect fit and opening for services should be sealed. Fire-stopping delays the spread of fire and, generally, the spread of smoke as well.

Openings for pipes

9.2 Pipes passing through a fire-separating element, unless in a protected shaft, should meet one of the alternatives A, B or C below.

Alternative A: Proprietary seals (any pipe diameter)

9.3 Provide a proprietary, tested sealing system that will maintain the fire resistance of the wall, floor or cavity barrier.

Alternative B: Pipes with a restricted diameter

9.4 Where a proprietary sealing system is not used, fire-stop around the pipe, keeping the opening for the pipe as small as possible. The nominal internal diameter of the pipe should not exceed the relevant dimension given in Table 9.1. The diameter given in Table 9.1 for pipes of specification (b) used in situation 2 or 3 assumes that the pipes are part of an above-ground drainage system and are enclosed as shown in Diagram 9.1. If they are not, the smaller diameter given for situation 5 should be used.

Alternative C: Sleeving

9.5 A pipe with a maximum nominal internal diameter of 160mm may be used with a sleeve made out of a high melting point metal, as shown in Diagram 9.2, if the pipe is made of one of the following.

a. Lead.

b. Aluminium.

c. Aluminium alloy.

d. Fibre-cement.

e. uPVC (pipes should also comply with either **BS 4514** or **BS 5255**).

A high melting point metal means any metal (such as cast iron, copper or steel) which, if exposed to a temperature of 800°C, will not soften or fracture to the extent that flame or hot gas will pass through the wall of the pipe.

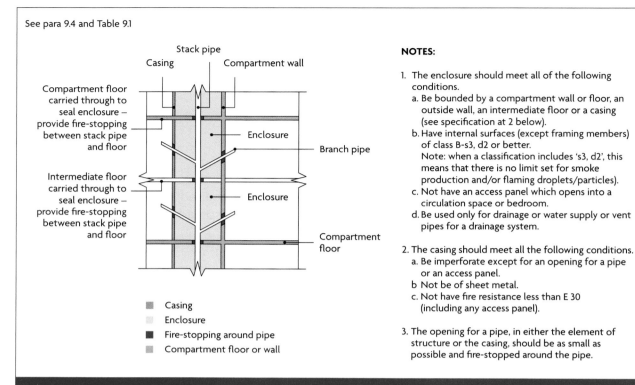

See para 9.4 and Table 9.1

NOTES:

1. The enclosure should meet all of the following conditions.
   a. Be bounded by a compartment wall or floor, an outside wall, an intermediate floor or a casing (see specification at 2 below).
   b. Have internal surfaces (except framing members) of class B-s3, d2 or better.
   Note: when a classification includes 's3, d2', this means that there is no limit set for smoke production and/or flaming droplets/particles).
   c. Not have an access panel which opens into a circulation space or bedroom.
   d. Be used only for drainage or water supply or vent pipes for a drainage system.

2. The casing should meet all the following conditions.
   a. Be imperforate except for an opening for a pipe or an access panel.
   b Not be of sheet metal.
   c. Not have fire resistance less than E 30 (including any access panel).

3. The opening for a pipe, in either the element of structure or the casing, should be as small as possible and fire-stopped around the pipe.

Diagram 9.1   Enclosure for drainage or water supply pipes

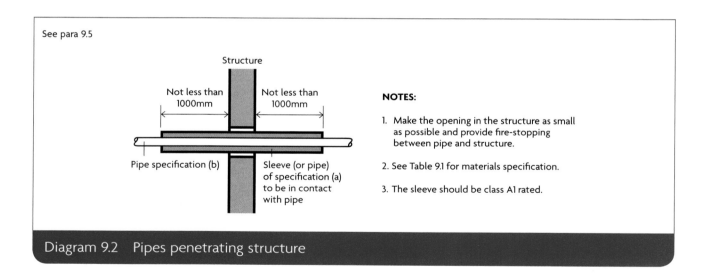

See para 9.5

NOTES:

1. Make the opening in the structure as small as possible and provide fire-stopping between pipe and structure.

2. See Table 9.1 for materials specification.

3. The sleeve should be class A1 rated.

Diagram 9.2   Pipes penetrating structure

Table 9.1 Maximum nominal internal diameter of pipes passing through a compartment wall/floor

Situation	Pipe material and maximum nominal internal diameter (mm)		
	(a) High melting point metal[1]	(b) Lead, aluminium, aluminium alloy, uPVC,[2] fibre cement	(c) Any other material
1. Structure (but not a wall separating buildings) enclosing a protected shaft that is not a stair or a lift shaft	160	110	40
2. Compartment wall or compartment floor between flats	160	160 (stack pipe)[3] 110 (branch pipe)[3]	40
3. Wall separating dwellinghouses	160	160 (stack pipe)[3] 110 (branch pipe)[3]	40
4. Wall or floor separating a dwellinghouse from an attached garage	160	110	40
5. Any other situation	160	40	40

NOTES:

1. Any metal (such as cast iron, copper or steel) which, if exposed to a temperature of 800°C, will not soften or fracture to the extent that flame or hot gas will pass through the wall of the pipe.

2. uPVC pipes that comply with either **BS 4514** or **BS 5255**.

3. These diameters are only in relation to pipes that form part of an above-ground drainage system and are enclosed as shown in Diagram 9.1. In other cases, the maximum diameters given for situation 5 apply.

Mechanical ventilation and air-conditioning systems

General provisions

9.6 Ductwork should not help to transfer fire and smoke through the building. Terminals of exhaust points should be sited away from final exits, cladding or roofing materials achieving class B-s3, d2 or worse and openings into the building.

9.7 Ventilation ducts supplying or extracting air directly to or from a protected stairway should not also serve other areas. A separate ventilation system should be provided for each protected stairway.

9.8 A fire and smoke damper should be provided where ductwork enters or leaves each section of the protected escape route it serves. It should be operated by a smoke detector or suitable fire detection system. Fire and smoke dampers should close when smoke is detected. Alternatively, the methods set out in paragraphs 9.16 and 9.17 and Diagrams 9.3 and 9.4 may be followed.

9.9 In a system that recirculates air, smoke detectors should be fitted in the extract ductwork before both of the following.

a. The point where recirculated air is separated from air to be discharged to the outside.

b. Any filters or other air cleaning equipment.

   When smoke is detected, detectors should do one of the following.

   i. Cause the system to immediately shut down.

ii. Switch the ventilation system from recirculating mode to extraction to divert smoke to outside the building.

9.10 In mixed use buildings, non-domestic kitchens, car parks and plant rooms should have separate and independent extraction systems. Extracted air should not be recirculated.

9.11 Under fire conditions, ventilation and air-conditioning systems should be compatible with smoke control systems and need to be considered in their respective design.

Ventilation ducts and flues passing through fire-separating elements

General provisions

9.12 If air handling ducts pass through fire-separating elements, the load-bearing capacity, integrity and insulation of the elements should be maintained using one or more of the following four methods. In most ductwork systems, a combination of the four methods is best to combat potential fire dangers.

a. Method 1 – thermally activated fire dampers.

b. Method 2 – fire resisting enclosures.

c. Method 3 – protection using fire resisting ductwork.

d. Method 4 – automatically activated fire and smoke dampers triggered by smoke detectors.

9.13 Further information on fire resisting ductwork is given in the ASFP Blue Book.

Flats and dwellings

9.14 Where ducts pass between fire-separating elements to serve multiple flats or dwellings, fire dampers or fire and smoke dampers should be actuated by both of the following.

a. Smoke detector-controlled automatic release mechanisms.

b. Thermally actuated devices.

Kitchen extract

9.15 Methods 1 and 4 should not be used for extract ductwork serving kitchens. The likely build-up of grease within the duct can adversely affect dampers.

Ducts passing through protected escape routes

9.16 Method 1 should not be used for extract ductwork passing through the enclosures of protected escape routes (Diagrams 9.3 and 9.4), as large volumes of smoke can pass thermal devices without triggering them.

9.17 An ES classified fire and smoke damper which is activated by a suitable fire detection system (method 4) may also be used for protected escape routes.

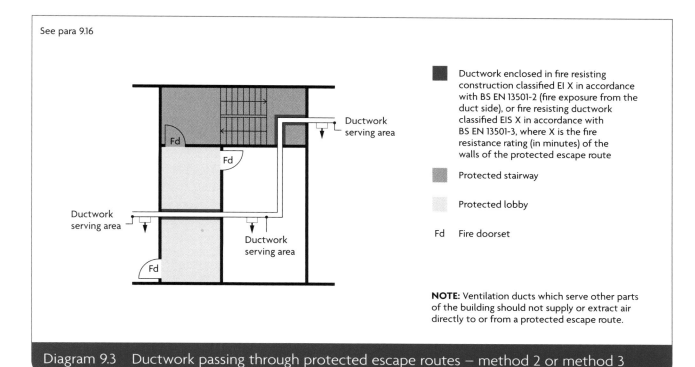

See para 9.16

Ductwork serving area

Ductwork serving area

Ductwork serving area

Fd

Fd

Fd

Ductwork enclosed in fire resisting construction classified EI X in accordance with BS EN 13501-2 (fire exposure from the duct side), or fire resisting ductwork classified EIS X in accordance with BS EN 13501-3, where X is the fire resistance rating (in minutes) of the walls of the protected escape route

Protected stairway

Protected lobby

Fd   Fire doorset

NOTE: Ventilation ducts which serve other parts of the building should not supply or extract air directly to or from a protected escape route.

Diagram 9.3   Ductwork passing through protected escape routes – method 2 or method 3

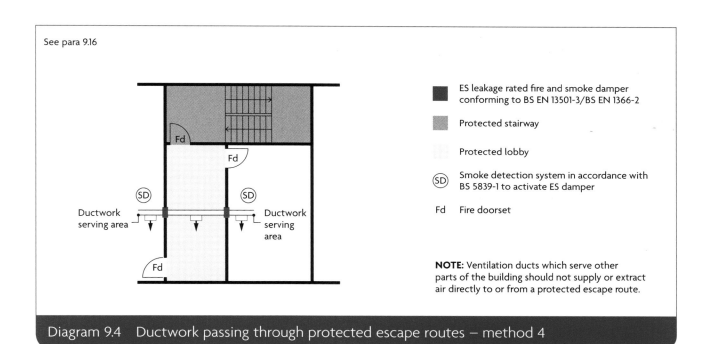

See para 9.16

SD

SD

Ductwork serving area

Ductwork serving area

Fd

Fd

Fd

ES leakage rated fire and smoke damper conforming to BS EN 13501-3/BS EN 1366-2

Protected stairway

Protected lobby

(SD) Smoke detection system in accordance with BS 5839-1 to activate ES damper

Fd   Fire doorset

NOTE: Ventilation ducts which serve other parts of the building should not supply or extract air directly to or from a protected escape route.

Diagram 9.4   Ductwork passing through protected escape routes – method 4

B3

Installation and specification of fire dampers

9.18 Both fire dampers and fire and smoke dampers should be all of the following.

a. Sited within the thickness of the fire-separating elements.

b. Securely fixed.

c. Sited such that, in a fire, expansion of the ductwork would not push the fire damper through the structure.

9.19 Access to the fire damper and its actuating mechanism should be provided for inspection, testing and maintenance.

9.20 Fire dampers should meet both of the following conditions.

a. Conform to **BS EN 15650**.

b. Have a minimum E classification of 60 minutes or to match the integrity rating of the fire resisting elements, whichever is higher.

9.21 Fire and smoke dampers should meet both of the following conditions.

a. Conform to **BS EN 15650**.

b. Have a minimum ES classification of 60 minutes or to match the integrity rating of the fire resisting elements, whichever is higher.

9.22 Smoke detectors should be sited so as to prevent the spread of smoke as early as practicable by activating the fire and smoke dampers. Smoke detectors and automatic release mechanisms used to activate fire dampers and/or fire and smoke dampers should conform to **BS EN 54-7** and **BS 5839-3** respectively.

Further information on fire dampers and/or fire and smoke dampers is given in the ASFP Grey Book.

Flues, etc.

9.23 The wall of a flue, duct containing flues or appliance ventilation duct(s) should have a fire resistance (REI) that is at least half of any compartment wall or compartment floor it passes through or is built into (Diagram 9.5).

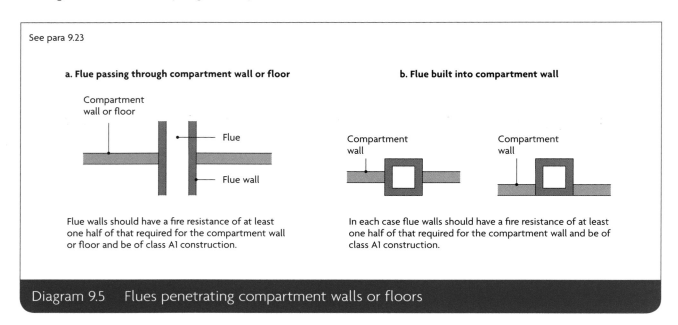

Diagram 9.5   Flues penetrating compartment walls or floors

Fire-stopping

9.24 In addition to any other provisions in this section, both of the following conditions should be met.

    a. Joints between fire-separating elements should be fire-stopped.

    b. Openings through a fire resisting element for pipes, ducts, conduits or cable should be all of the following.

        i. As few as possible.

        ii. As small as practicable.

        iii. Fire-stopped (allowing thermal movement in the case of a pipe or duct).

    NOTE: The fire-stopping around fire dampers, fire resisting ducts, fire and smoke dampers and smoke control ducts should be in accordance with the manufacturer or supplier's installation instructions.

9.25 Materials used for fire-stopping should be reinforced with (or supported by) materials rated class A2-s3, d2 or better to prevent displacement in both of the following cases.

    a. Where the unsupported span is greater than 100mm.

    b. Where non-rigid materials are used (unless subjected to appropriate fire resistance testing to show their suitability).

9.26 Proprietary, tested fire-stopping and sealing systems are available and may be used. Different materials suit different situations and not all are suitable in every situation.

9.27 Other fire-stopping materials include the following.

    a. Cement mortar.

    b. Gypsum-based plaster.

    c. Cement-based or gypsum-based vermiculite/perlite mixes.

    d. Glass fibre, crushed rock, blast furnace slag or ceramic-based products (with or without resin binders).

    e. Intumescent mastics.

These may be used in situations appropriate to the particular material. Not all materials will be suitable in every situation.

9.28 Guidance on the design, installation and maintenance of measures to contain fires or slow their spread is given in *Ensuring Best Practice for Passive Fire Protection in Buildings* produced by the Association for Specialist Fire Protection (ASFP).

9.29 Further information on generic systems, their suitability for different applications and guidance on test methods, is given in the ASFP Red Book.

Requirement B4: External fire spread

These sections deal with the following requirement from Part B of Schedule 1 to the Building Regulations 2010. Section 10 also refers to regulation 7(2) of the Building Regulations 2010. Guidance on regulation 7(1) can be found in Approved Document 7.

Requirement

Requirement   *Limits on application*

External fire spread

B4. (1) The external walls of the building shall adequately resist the spread of fire over the walls and from one building to another, having regard to the height, use and position of the building.

(2) The roof of the building shall adequately resist the spread of fire over the roof and from one building to another, having regard to the use and position of the building.

Regulation

Regulation 7 – Materials and workmanship

(1) Building work shall be carried out—

(a) with adequate and proper materials which—

(i) are appropriate for the circumstances in which they are used,

(ii) are adequately mixed or prepared, and

(iii) are applied, used or fixed so as adequately to perform the functions for which they are designed; and

(b) in a workmanlike manner.

(2) Subject to paragraph (3), building work shall be carried out so that materials which become part of an external wall, or specified attachment, of a relevant building are of European Classification A2-s1, d0 or A1, classified in accordance with BS EN 13501-1:2007+A1:2009 entitled "Fire classification of construction products and building elements. Classification using test data from reaction to fire tests" (ISBN 978 0 580 59861 6) published by the British Standards Institution on 30th March 2007 and amended in November 2009.

Regulation *continued*

(3) Paragraph (2) does not apply to—

    (a) cavity trays when used between two leaves of masonry;

    (b) any part of a roof (other than any part of a roof which falls within paragraph (iv) of regulation 2(6)) if that part is connected to an external wall;

    (c) door frames and doors;

    (d) electrical installations;

    (e) insulation and water proofing materials used below ground level;

    (f) intumescent and fire stopping materials where the inclusion of the materials is necessary to meet the requirements of Part B of Schedule 1;

    (g) membranes;

    (h) seals, gaskets, fixings, sealants and backer rods;

    (i) thermal break materials where the inclusion of the materials is necessary to meet the thermal bridging requirements of Part L of Schedule 1; or

    (j) window frames and glass.

(4) In this regulation—

    (a) a "relevant building" means a building with a storey (not including roof-top plant areas or any storey consisting exclusively of plant rooms) at least 18 metres above ground level and which—

        (i) contains one or more dwellings;

        (ii) contains an institution; or

        (iii) contains a room for residential purposes (excluding any room in a hostel, hotel or boarding house); and

    (b) "above ground level" in relation to a storey means above ground level when measured from the lowest ground level adjoining the outside of a building to the top of the floor surface of the storey.

Intention

Resisting fire spread over external walls

The external envelope of a building should not contribute to undue fire spread from one part of a building to another part. This intention can be met by constructing external walls so that both of the following are satisfied.

a. The risk of ignition by an external source to the outside surface of the building and spread of fire over the outside surface is restricted.

b. The materials used to construct external walls, and attachments to them, and how they are assembled do not contribute to the rate of fire spread up the outside of the building.

The extent to which this is necessary depends on the height and use of the building.

Resisting fire spread from one building to another

The external envelope of a building should not provide a medium for undue fire spread to adjacent buildings or be readily ignited by fires in adjacent buildings. This intention can be met by constructing external walls so that all of the following are satisfied.

a. The risk of ignition by an external source to the outside surface of the building is restricted.

b. The amount of thermal radiation that falls on a neighbouring building from window openings and other unprotected areas in the building on fire is not enough to start a fire in the other building.

c. Flame spread over the roof and/or fire penetration from external sources through the roof is restricted.

The extent to which this is necessary depends on the use of the building and its position in relation to adjacent buildings and therefore the site boundary.

Section 10: Resisting fire spread over external walls

Introduction

10.1 The external wall of a building should not provide a medium for fire spread if that is likely to be a risk to health and safety. Combustible materials and cavities in external walls and attachments to them can present such a risk, particularly in tall buildings. The guidance in this section is designed to reduce the risk of vertical fire spread as well as the risk of ignition from flames coming from adjacent buildings.

Fire resistance

10.2 This section does not deal with fire resistance for external walls. An external wall may need fire resistance to meet the requirements of Section 3 (Means of escape – flats), Section 6 (Loadbearing elements of structures – flats) or Section 11 (Resisting fire spread from one building to another).

Combustibility of external walls

10.3 The external walls of buildings other than those described in regulation 7(4) of the Building Regulations should achieve either of the following.

a. Follow the provisions given in paragraphs 10.5 to 10.8, which provide guidance on all of the following.

    i.   External surfaces.

    ii.   Materials and products.

    iii. Cavities and cavity barriers.

b. Meet the performance criteria given in BRE report BR 135 for external walls using full-scale test data from **BS 8414-1** or **BS 8414-2**.

10.4 In relation to buildings of any height or use, consideration should be given to the choice of materials (including their extent and arrangement) used for the external wall, or attachments to the wall, to reduce the risk of fire spread over the wall.

External surfaces

10.5 The external surfaces (i.e. outermost external material) of external walls should comply with the provisions in Table 10.1. The provisions in Table 10.1 apply to each wall individually in relation to its proximity to the relevant boundary.

Table 10.1   Reaction to fire performance of external surface of walls

Building type	Building height	Less than 1000mm from the relevant boundary	1000mm or more from the relevant boundary
'Relevant buildings' as defined in regulation 7(4) (see paragraph 10.10)		Class A2-s1, d0[1] or better	Class A2-s1, d0[1] or better
Assembly and recreation	More than 18m	Class B-s3, d2[2] or better	From ground level to 18m: class C-s3, d2[3] or better
			From 18m in height and above: class B-s3, d2[2] or better
	18m or less	Class B-s3, d2[2] or better	Up to 10m above ground level: class C-s3, d2[3] or better
			Up to 10m above a roof or any part of the building to which the public have access: class C-s3, d2[3] or better[4]
			From 10m in height and above: no minimum performance
Any other building	More than 18m	Class B-s3, d2[2] or better	From ground level to 18m: class C-s3, d2[3] or better
			From 18m in height and above: class B-s3, d2[2] or better
	18m or less	Class B-s3, d2[2] or better	No provisions

NOTES:

In addition to the requirements within this table, buildings with a top occupied storey above 18m should also meet the provisions of paragraph 10.6.

In all cases, the advice in paragraph 10.4 should be followed.

1. The restrictions for these buildings apply to all the materials used in the external wall and specified attachments (see paragraphs 10.9 to 10.12 for further guidance).

2. Profiled or flat steel sheet at least 0.5 mm thick with an organic coating of no more than 0.2mm thickness is also acceptable.

3. Timber cladding at least 9mm thick is also acceptable.

4. 10m is measured from the top surface of the roof.

Materials and products

10.6 In a building with a storey 18m or more in height (see Diagram D6 in Appendix D) any insulation product, filler material (such as the core materials of metal composite panels, sandwich panels and window spandrel panels but not including gaskets, sealants and similar) etc. used in the construction of an external wall should be class A2-s3, d2 or better (see Appendix B). This restriction does not apply to masonry cavity wall construction which complies with Diagram 8.2 in Section 8. Where regulation 7(2) applies, that regulation prevails over all the provisions in this paragraph.

10.7 Best practice guidance for green walls (also called living walls) can be found in *Fire Performance of Green Roofs and Walls*, published by the Department for Communities and Local Government.

Cavities and cavity barriers

10.8 Cavity barriers should be provided in accordance with Section 5 in dwellinghouses and Section 8 in flats.

Regulation 7(2) and requirement B4

Materials

10.9 Regulation 7(1)(a) requires that materials used in building work are appropriate for the circumstances in which they are used. Regulation 7(2) sets requirements in respect of external walls and specified attachments in relevant buildings.

> **NOTE:** Guidance on regulation 7(1) can be found in Approved Document 7.

10.10 Regulation 7(2) applies to any building with a storey at least 18m above ground level (as measured in accordance with Diagram D6 in Appendix D) and which contains one or more dwellings; an institution; or a room for residential purposes (excluding any room in a hostel, hotel or a boarding house). It requires that all materials which become part of an external wall or specified attachment achieve class A2-s1, d0 or class A1, other than those exempted by regulation 7(3).

> **NOTE:** The above includes student accommodation, care homes, sheltered housing, hospitals and dormitories in boarding schools. See regulation 7(4) for the definition of relevant buildings.

> **NOTE:** The requirement in regulation 7(2) is limited to materials achieving class A2-s1, d0 or class A1.

10.11 External walls and specified attachments are defined in regulation 2 and these definitions include any parts of the external wall as well as balconies, solar panels and sun shading.

10.12 Regulation 7(3) provides an exemption for certain components found in external walls and specified attachments.

Material change of use

10.13 Regulations 5(k) and 6(3) provide that, where the use of a building is changed such that the building becomes a building described in regulation 7(4), the construction of the external walls, and specified attachments, must be investigated and, where necessary, work must be carried out to ensure they only contain materials achieving class A2-s1, d0 or class A1, other than those exempted by regulation 7(3).

Additional considerations

10.14 The provisions of regulation 7 apply in addition to requirement B4. Therefore, for buildings described in regulation 7(4), the potential impact of any products incorporated into or onto the external walls and specified attachments should be carefully considered with regard to their number, size, orientation and position.

10.15 Particular attention is drawn to the following points.

a. Membranes used as part of the external wall construction above ground level should achieve a minimum of class B-s3, d0.

b. Internal linings should comply with the guidance provided in Section 4.

c. Any part of a roof should achieve the minimum performance as detailed in Section 12.

d. As per regulation 7(3), window frames and glass (including laminated glass) are exempted from regulation 7(2). Window spandrel panels and infill panels must comply with regulation 7(2).

e. Thermal breaks are small elements used as part of the external wall construction to restrict thermal bridging. There is no minimum performance for these materials. However, they should not span two compartments and should be limited in size to the minimum required to restrict the thermal bridging (the principal insulation layer is not to be regarded as a thermal break).

f. Regulation 7(2) only applies to specified attachments. Shop front signs and similar attachments are not covered by the requirements of regulation 7(2), although attention is drawn to paragraph 10.15g.

g. While regulation 7(2) applies to materials which become part of an external wall or specified attachment, consideration should be given to other attachments to the wall which could impact on the risk of fire spread over the wall.

Section 11: Resisting fire spread from one building to another

Introduction

11.1 The following assumptions enable a reasonable standard of resistance to the spread of fire to be specified.

    a. The size of a fire depends on the compartmentation within the building. A fire may involve a complete compartment, but will not spread to other compartments.

    b. The intensity of a fire is related to the building use, but can be moderated by a sprinkler system.

    c. Fires in 'residential' and 'assembly and recreation' buildings (purpose groups 1, 2 and 5) represent a greater risk to life.

    d. A building on the far side of the relevant boundary meets both of the following conditions.

        i. Has a similar elevation to the one in question.

        ii. Is the same distance as the one in question from the common boundary.

    e. The radiated heat passing through any part of the fire resisting external wall may be discounted.

11.2 Where regulation 7(2) applies, that regulation prevails over the provisions within this section.

11.3 If a reduced separation distance between buildings, or increased amount of unprotected area, is required, smaller compartments should be considered.

Boundaries

11.4 The fire resistance of a wall depends on its distance from the relevant boundary (see Diagram 11.1). Separation distances are measured to boundaries to ensure that the location and design of buildings on adjoining sites have no influence on the building under consideration.

11.5 The boundary that a wall faces is the relevant boundary (Diagram 11.2). It may be one of the following.

    a. The site boundary.

    b. The centre line of a space where further development is unlikely, such as a road, railway, canal or river.

    c. An assumed notional boundary between two buildings on the same site (Diagram 11.3) where either of the following conditions is met.

        i. One or both of the buildings are in the 'residential' or 'assembly and recreation' purpose groups (purpose group 1 or 5).

        ii. The buildings will be operated/managed by different organisations.

See para 11.4

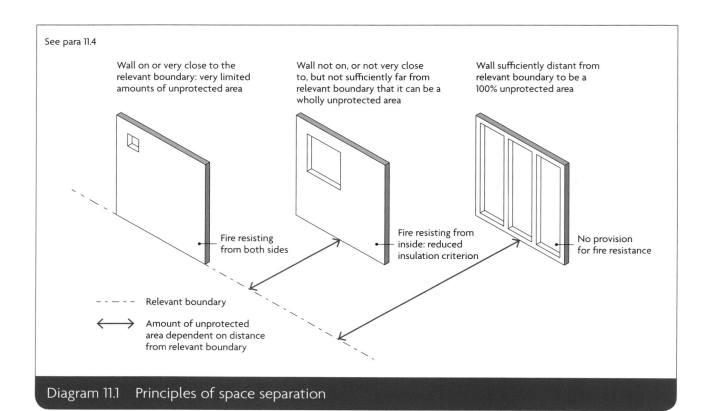

Wall on or very close to the relevant boundary: very limited amounts of unprotected area

Wall not on, or not very close to, but not sufficiently far from relevant boundary that it can be a wholly unprotected area

Wall sufficiently distant from relevant boundary to be a 100% unprotected area

Fire resisting from both sides

Fire resisting from inside: reduced insulation criterion

No provision for fire resistance

– · – · – · Relevant boundary

⟷ Amount of unprotected area dependent on distance from relevant boundary

Diagram 11.1   Principles of space separation

See para 11.5

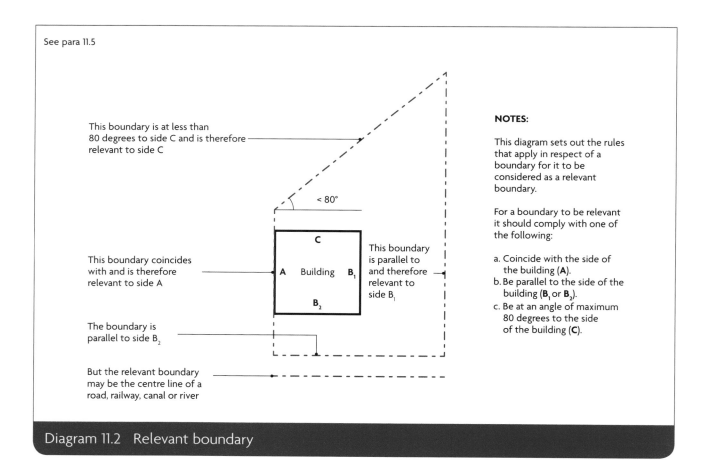

This boundary is at less than 80 degrees to side C and is therefore relevant to side C

< 80°

This boundary coincides with and is therefore relevant to side A

This boundary is parallel to and therefore relevant to side B₁

The boundary is parallel to side B₂

But the relevant boundary may be the centre line of a road, railway, canal or river

NOTES:

This diagram sets out the rules that apply in respect of a boundary for it to be considered as a relevant boundary.

For a boundary to be relevant it should comply with one of the following:

a. Coincide with the side of the building (**A**).
b. Be parallel to the side of the building (**B₁** or **B₂**).
c. Be at an angle of maximum 80 degrees to the side of the building (**C**).

Diagram 11.2   Relevant boundary

See para 11.5

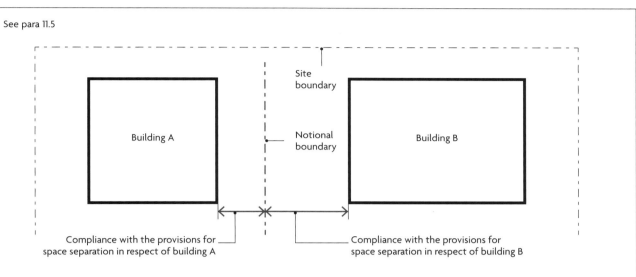

Compliance with the provisions for space separation in respect of building A

Compliance with the provisions for space separation in respect of building B

NOTES:

The notional boundary should be set in the area between the two buildings using the following rules:

1. The notional boundary is assumed to exist in the space between the buildings and is positioned so that one of the buildings would comply with the provisions for space separation having regard to the amount of its unprotected area. In practice, if one of the buildings is existing, the position of the boundary will be set by the space separation factors for that building.

2. The siting of the new building, or the second building if both are new, can then be checked to see that it also complies, using the notional boundary as the relevant boundary for the second building.

Diagram 11.3   Notional boundary

Unprotected areas and fire resistance

11.6 Parts of an external wall with less fire resistance than the appropriate amount given in Appendix B, Table B4, are called unprotected areas.

11.7 Where a fire resisting external wall has a surface material that is worse than class B-s3, d2 and is more than 1mm thick, that part of the wall should be classified as an unprotected area equating to half its area (Diagram 11.4).

External walls on, and within 1000mm of, the relevant boundary

11.8 Unprotected areas should meet the conditions in Diagram 11.5, and the rest of the wall should be fire resisting from both sides.

External surface materials facing the boundary should be class B-s3, d2 or better.

External walls 1000mm or more from the relevant boundary

11.9 Unprotected areas should not exceed the result given by one of the methods in paragraph 11.16, and the rest of the wall (if any) should be fire resisting but only from the inside of the building.

External walls of protected stairways

11.10 Exclude external walls of protected stairways when assessing unprotected areas (see Diagram 3.10).

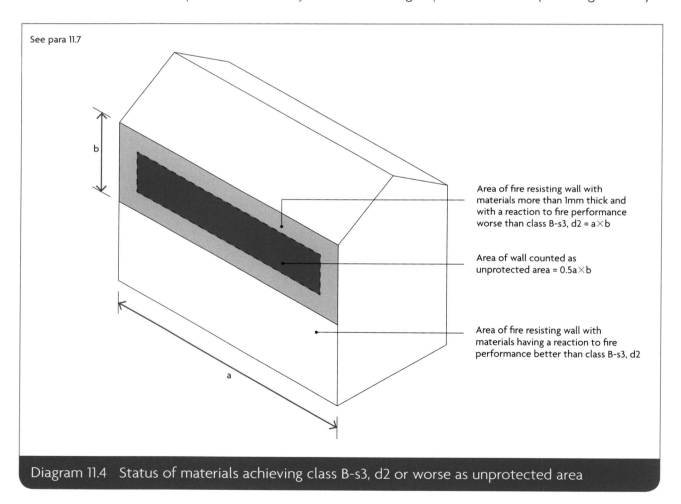

See para 11.7

b

a

Area of fire resisting wall with materials more than 1mm thick and with a reaction to fire performance worse than class B-s3, d2 = a×b

Area of wall counted as unprotected area = 0.5a×b

Area of fire resisting wall with materials having a reaction to fire performance better than class B-s3, d2

Diagram 11.4   Status of materials achieving class B-s3, d2 or worse as unprotected area

Small unprotected areas

11.11 In an otherwise protected wall, small unprotected areas may be ignored where they meet the conditions in Diagram 11.5.

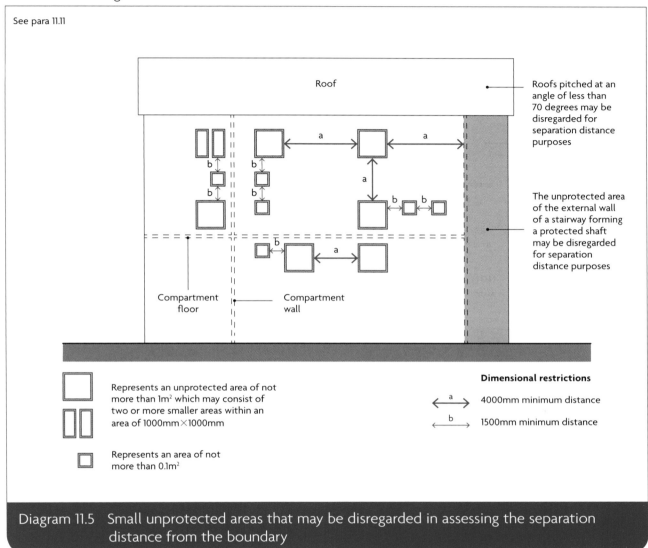

See para 11.11

Roof

Roofs pitched at an angle of less than 70 degrees may be disregarded for separation distance purposes

The unprotected area of the external wall of a stairway forming a protected shaft may be disregarded for separation distance purposes

Compartment floor

Compartment wall

☐ Represents an unprotected area of not more than 1m² which may consist of two or more smaller areas within an area of 1000mm×1000mm

☐ Represents an area of not more than 0.1m²

Dimensional restrictions

a — 4000mm minimum distance

b — 1500mm minimum distance

Diagram 11.5 Small unprotected areas that may be disregarded in assessing the separation distance from the boundary

Canopies

11.12 Where both of the following apply, separation distances may be determined from the wall rather than from the edge of the canopy (Diagram 11.6).

a. The canopy is attached to the side of a building.

b. The edges of the canopy are a minimum of 2m from the relevant boundary.

Canopies that fall within class 6 or class 7 of Schedule 2 to the regulations (Exempt Buildings and Work) are exempt from the Building Regulations.

11.13 Space separation may be disregarded if a canopy is all of the following.

a. Free-standing.

b. Above a limited risk or controlled hazard.

c. A minimum of 1000mm from the relevant boundary.

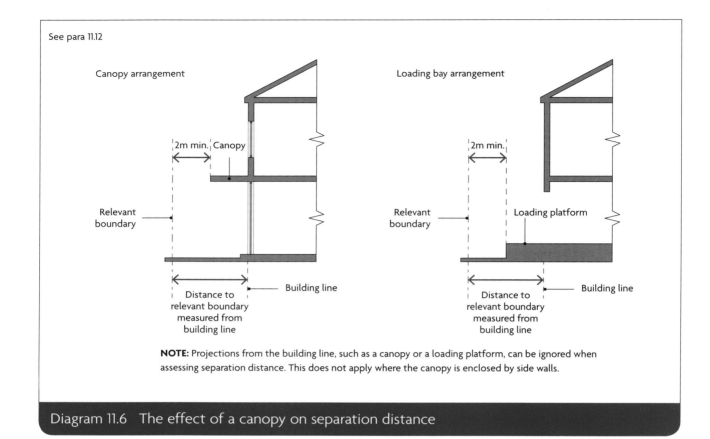

See para 11.12

Canopy arrangement

2m min. Canopy

Relevant
boundary

Building line

Distance to
relevant boundary
measured from
building line

Loading bay arrangement

2m min.

Relevant
boundary

Loading platform

Building line

Distance to
relevant boundary
measured from
building line

NOTE: Projections from the building line, such as a canopy or a loading platform, can be ignored when assessing separation distance. This does not apply where the canopy is enclosed by side walls.

Diagram 11.6   The effect of a canopy on separation distance

Roofs

11.14 Roofs with a pitch of more than 70 degrees to the horizontal should be assessed in accordance with this section. Vertical parts of a pitched roof, such as dormer windows, should be included *only* if the slope of the roof exceeds 70 degrees.

It is a matter of judgement whether a continuous run of dormer windows that occupies most of a steeply pitched roof should be treated as a wall rather than a roof.

Portal frames

11.15 Portal frames are often used in single storey industrial and commercial buildings where there may be no need for fire resistance of the structure (requirement B3). However, where a portal framed building is near a relevant boundary, the external wall near the boundary may need fire resistance to restrict the spread of fire between buildings. It is generally accepted that a portal frame acts as a single structural element because of the moment-resisting connections used, especially at the column/rafter joints. Thus, in cases where the external wall of the building cannot be wholly unprotected, the rafter members of the frame, as well as the column members, may need to be fire protected. The design method for this is set out in SCI Publication P313.

NOTE: The recommendations in the SCI publication for designing the foundation to resist overturning do not need to be followed if the building is fitted with a sprinkler system in accordance with Appendix E.

NOTE: Normally, portal frames of reinforced concrete can support external walls requiring a similar degree of fire resistance without specific provision at the base to resist overturning.

NOTE: Existing buildings may have been designed to comply with all of the following guidance, which is also acceptable.

a. The column members are fixed rigidly to a base of sufficient size and depth to resist overturning.

b. There is brick, block or concrete protection to the columns up to a protected ring beam providing lateral support.

c. There is some form of roof venting to give early heat release. (The roof venting could be, for example, PVC rooflights covering some 10% of the floor area and evenly spaced over the floor area.)

Methods for calculating acceptable unprotected area

11.16 Two simple methods are given for calculating the acceptable amount of unprotected area in an external wall that is a minimum of 1000mm from any point on the relevant boundary. More precise methods are described in BRE report BR 187 and may be used instead.

Method 1

11.17 This method applies to small buildings intended to be used for blocks of flats or dwellinghouses.

11.18 The building should not exceed three storeys in height (excluding basements) or 24m in length. Each side of the building should meet the limits stated in Diagram 11.7. Any small unprotected areas falling within the limits shown in Diagram 11.5 can be ignored.

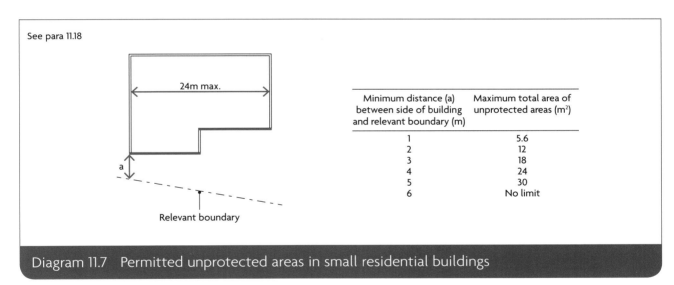

See para 11.18

24m max.

a

Relevant boundary

Minimum distance (a) between side of building and relevant boundary (m)	Maximum total area of unprotected areas (m²)
1	5.6
2	12
3	18
4	24
5	30
6	No limit

Diagram 11.7   Permitted unprotected areas in small residential buildings

Method 2

11.19 This method may be used for buildings or compartments for which method 1 is not appropriate.

11.20 The building should not exceed 10m in height. Each side of the building should meet the limits stated in Table 11.1. Any areas falling within the limits shown in Diagram 11.5 can be ignored.

Table 11.1 Permitted unprotected areas in small buildings or compartments

Minimum distance between side of building and relevant boundary (m)	Maximum total percentage of unprotected area (%)[1]
Not applicable	4
1	8
2.5	20
5	40
7.5	60
10	80
12.5	100

NOTES:

Intermediate values may be obtained by interpolation.

1. The total percentage of unprotected area is found by dividing the total unprotected area by the area of a rectangle that encloses all the unprotected areas, and multiplying the result by 100.

Sprinkler systems

11.21 If a building is fitted throughout with a sprinkler system in accordance with Appendix E, either of the following is permitted.

a. The boundary distance can be halved, to a minimum distance of 1m.

b. The amount of unprotected area can be doubled.

Section 12: Resisting fire spread over roof coverings

Introduction

12.1 'Roof covering' describes one or more layers of material, but not the roof structure as a whole.

12.2 Provisions for the fire properties of roofs are given in other parts of this document.

    a. Requirement B1 – for roofs that are part of a means of escape.

    b. Requirement B2 – for the internal surfaces of rooflights as part of internal linings.

    c. Requirement B3 – for roofs that are used as a floor and for roofs passing over a compartment wall.

    d. Section 11 – the circumstances in which a roof is subject to the provisions for space separation.

Separation distances

12.3 Separation distance is the minimum distance from the roof, or part of the roof, to the relevant boundary (paragraph 11.4). Table 12.1 sets out separation distances by the type of roof covering and the size and use of the building.

In addition, roof covering products (and/or materials) defined in Commission Decision 2000/553/EC of 6 September 2000, implementing Council Directive 89/106/EEC, can be considered to fulfil all of the requirements for the performance characteristic 'external fire performance' without the need for testing, *provided that any national provisions on the design and execution of works are fulfilled*, and can be used without restriction.

12.4 The performance of rooflights is specified in a similar way to the performance of roof coverings. Plastic rooflights may also be used.

Plastic rooflights

12.5 Table 12.2 and Diagram 12.1 set the limitations for using plastic rooflights whose lower surface has a minimum class D-s3, d2 rating.

12.6 Table 12.3 sets the limitations for using thermoplastic materials with a TP(a) rigid or TP(b) (see also Diagram 12.1) classification. The method of classifying thermoplastic materials is given in Appendix B.

12.7 Other than for the purposes of Diagram 5.2, polycarbonate or uPVC rooflights achieving a minimum rating of class C-s3, d2 can be regarded as having a $B_{ROOF}(t4)$ classification.

Unwired glass in rooflights

12.8 When used in rooflights, unwired glass a minimum of 4mm thick can be regarded as having a B_{ROOF}(t4) classification.

Thatch and wood shingles

12.9 If the performance of thatch or wood shingles cannot be established, they should be regarded as having an E_{ROOF}(t4) classification in Table 12.1.

NOTE: Consideration can be given to thatched roofs being closer to the relevant boundary than shown in Table 12.1 if, for example, all of the following precautions (based on the LABC publication *Thatched Buildings (the Dorset Model): New Properties and Extensions*) are incorporated in the design.

a. The rafters are overdrawn with construction having not less than 30 minutes' fire resistance.

b. The guidance given in Approved Document J is followed.

c. The smoke alarm installation (see Section 1) extends to the roof spaces.

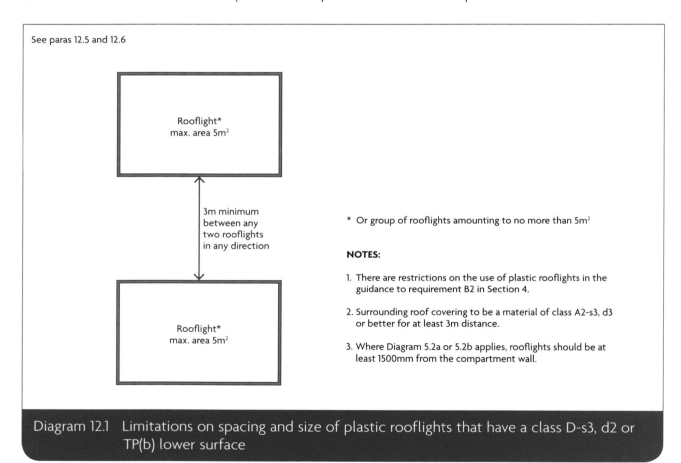

See paras 12.5 and 12.6

Rooflight*
max. area 5m²

3m minimum between any two rooflights in any direction

Rooflight*
max. area 5m²

* Or group of rooflights amounting to no more than 5m²

NOTES:

1. There are restrictions on the use of plastic rooflights in the guidance to requirement B2 in Section 4.

2. Surrounding roof covering to be a material of class A2-s3, d3 or better for at least 3m distance.

3. Where Diagram 5.2a or 5.2b applies, rooflights should be at least 1500mm from the compartment wall.

Diagram 12.1 Limitations on spacing and size of plastic rooflights that have a class D-s3, d2 or TP(b) lower surface

Table 12.1 Limitations on roof coverings

Designation[1] of covering of roof or part of roof	Distance from any point on relevant boundary			
	Less than 6m	At least 6m	At least 12m	At least 20m
B_{ROOF}(t4)	●	●	●	●
C_{ROOF}(t4)	○	●	●	●
D_{ROOF}(t4)	○	●[2][3]	●[2]	●
E_{ROOF}(t4)	○	●[2][3]	●[2]	●[2]
F_{ROOF}(t4)	○	○	○	●[2][3]

● Acceptable.

○ Not acceptable.

NOTES:

Separation distances do not apply to the boundary between roofs of a pair of semi-detached dwellinghouses and to enclosed/covered walkways. However, see Diagram 5.2 if the roof passes over the top of a compartment wall.

Polycarbonate and uPVC rooflights that achieve a class C-s3, d2 rating by test may be regarded as having a B_{ROOF}(t4) designation.

1. The designation of external roof surfaces is explained in Appendix B.

2. Not acceptable on any of the following buildings.

    a. Dwellinghouses in terraces of three or more dwellinghouses.

    b. Any other buildings with a cubic capacity of more than 1500m³.

3. Acceptable on buildings not listed in (1) if both of the following apply.

    a. Part of the roof has a maximum area of 3m² and is a minimum of 1500mm from any similar part.

    b. The roof between the parts is covered with a material rated class A2-s3, d2 or better.

Table 12.2 Class D-s3, d2 plastic rooflights: limitations on use and boundary distance

Minimum classification on lower surface[1]	Space that rooflight can serve	Minimum distance from any point on relevant boundary to rooflight with an external designation[2] of:	
		E_{ROOF}(t4) or D_{ROOF}(t4)	F_{ROOF}(t4)
Class D-s3, d2	a. Balcony, verandah, carport, covered way or loading bay with at least one longer side wholly or permanently open	6m	20m
	b. Detached swimming pool		
	c. Conservatory, garage or outbuilding, with a maximum floor area of 40m²		
	d. Circulation space[3] (except a protected stairway)	6m[4]	20m[4]
	e. Room[3]		

NOTES:

None of the above designations are suitable for protected stairways.

Polycarbonate and uPVC rooflights that achieve a class C-s3, d2 rating by test (see paragraph 12.7) may be regarded as having a B_{ROOF}(t4) classification.

Where Diagram 5.2a or 5.2b applies, rooflights should be a minimum of 1500mm from the compartment wall.

If double-skinned or laminate products have upper and lower surfaces of different materials, the greater distance applies.

1. See also the guidance to requirement B2 in Section 4.

2. The designation of external roof surfaces is explained in Appendix B.

3. Single-skinned rooflight only, in the case of non-thermoplastic material.

4. The rooflight should also meet the provisions of Diagram 12.1.

Table 12.3 TP(a) and TP(b) thermoplastic rooflights: limitations on use and boundary distance

Minimum classification on lower surface[1]	Space that rooflight can serve	Minimum distance from any point on relevant boundary to rooflight with an external designation[1] of:	
		TP(a)	TP(b)
1. TP(a) rigid	Any space except a protected stairway	6m[2]	Not applicable
2. TP(b)	a. Balcony, verandah, carport, covered way or loading bay with at least one longer side wholly or permanently open	Not applicable	6m
	b. Detached swimming pool		
	c. Conservatory, garage or outbuilding, with a maximum floor area of 40m²		
	d. Circulation space[3] (except a protected stairway)	Not applicable	6m[4]
	e. Room[3]		

NOTES:

None of the above designations are suitable for protected stairways.

Polycarbonate and uPVC rooflights that achieve a class C-s3, d2 rating by test (paragraph 12.7) may be regarded as having a B_{ROOF}(t4) classification.

Where Diagram 5.2a or 5.2b applies, rooflights should be a minimum of 1500mm from the compartment wall.

If double-skinned or laminate products have upper and lower surfaces of different materials, the greater distance applies.

1. See also the guidance to requirement B2 in section 4.

2. No limit in the case of any space described in 2a, b and c.

3. Single-skinned rooflight only, in the case of non-thermoplastic material.

4. The rooflight should also meet the provisions of diagram 12.1.

B5

Requirement B5: Access and facilities for the fire service

These sections deal with the following requirement from Part B of Schedule 1 to the Building Regulations 2010.

Requirement	
Requirement	*Limits on application*
Access and facilities for the fire service	
B5. (1) The building shall be designed and constructed so as to provide reasonable facilities to assist fire fighters in the protection of life.	
(2) Reasonable provision shall be made within the site of the building to enable fire appliances to gain access to the building.	

Intention

Provisions covering access and facilities for the fire service are to safeguard the health and safety of people in and around the building. Their extent depends on the size and use of the building. Most firefighting is carried out within the building. In the Secretary of State's view, requirement B5 is met by achieving all of the following.

a. External access enabling fire appliances to be used near the building.

b. Access into and within the building for firefighting personnel to both:

    i.   search for and rescue people

    ii.   fight fire.

c. Provision for internal fire facilities for firefighters to complete their tasks.

d. Ventilation of heat and smoke from a fire in a basement.

If an alternative approach is taken to providing the means of escape, outside the scope of this approved document, additional provisions for firefighting access may be required. Where deviating from the general guidance, it is advisable to seek advice from the fire and rescue service as early as possible (even if there is no statutory duty to consult).

Section 13: Vehicle access

Provision and design of access routes and hardstandings

13.1 For dwellinghouses, access for a pumping appliance should be provided to within 45m of all points inside the dwellinghouse.

13.2 For flats, either of the following provisions should be made.

 a. Provide access for a pumping appliance to within 45m of all points inside each flat of a block, measured along the route of the hose.

 b. Provide fire mains in accordance with paragraphs 13.5 and 13.6.

13.3 Access routes and hardstandings should comply with the guidance in Table 13.1.

13.4 Dead-end access routes longer than 20m require turning facilities, as in Diagram 13.1. Turning facilities should comply with the guidance in Table 13.1.

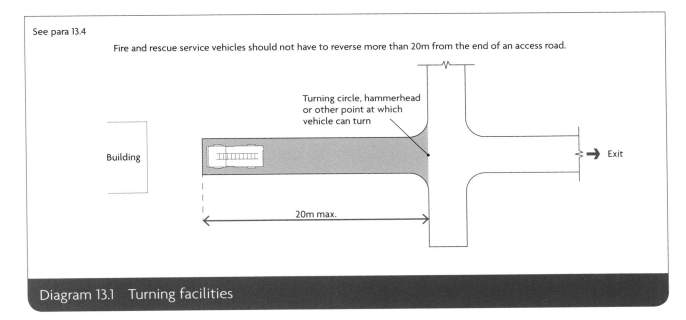

See para 13.4

Fire and rescue service vehicles should not have to reverse more than 20m from the end of an access road.

Turning circle, hammerhead or other point at which vehicle can turn

Building

Exit

20m max.

Diagram 13.1   Turning facilities

Table 13.1 Typical fire and rescue service vehicle access route specification

Appliance type	Minimum width of road between kerbs (m)	Minimum width of gateways (m)	Minimum turning circle between kerbs (m)	Minimum turning circle between walls (m)	Minimum clearance height (m)	Minimum carrying capacity (tonnes)
Pump	3.7	3.1	16.8	19.2	3.7	12.5
High reach	3.7	3.1	26.0	29.0	4.0	17.0

NOTES:

1. Fire appliances are not standardised. The building control body may, in consultation with the local fire and rescue service, use other dimensions.

2. The roadbase can be designed to 12.5 tonne capacity. Structures such as bridges should have the full 17-tonne capacity. The weight of high reach appliances is distributed over a number of axles, so infrequent use of a route designed to accommodate 12.5 tonnes should not cause damage.

Blocks of flats fitted with fire mains

13.5 For buildings fitted with dry fire mains, both of the following apply.

a. Access should be provided for a pumping appliance to within 18m of each fire main inlet connection point. Inlets should be on the face of the building.

b. The fire main inlet connection point should be visible from the parking position of the appliance, and satisfy paragraph 14.10.

13.6 For buildings fitted with wet fire mains, access for a pumping appliance should comply with both of the following.

a. Within 18m, and within sight, of an entrance giving access to the fire main.

b. Within sight of the inlet to replenish the suction tank for the fire main in an emergency.

Section 14: Fire mains and hydrants – flats

Introduction

14.1 Fire mains are installed for the fire and rescue service to connect hoses for water. They may be either of the following.

   a. The 'dry' type, which are both of the following.

      i. Normally kept empty.

      ii. Supplied through a hose from a fire and rescue service pumping appliance.

   b. The 'wet' type, which are both of the following.

      i. Kept full of water.

      ii. Supplied by pumps from tanks in the building.

There should be a facility to replenish a wet system from a pumping appliance in an emergency.

Provision of fire mains

14.2 Buildings with firefighting shafts should have fire mains provided in both of the following.

   a. The firefighting stairs.

   b. Where necessary, in protected stairways.

The criteria for providing firefighting shafts and fire mains are given in Section 15.

14.3 Buildings without firefighting shafts should be provided with fire mains where fire service vehicle access is not provided in accordance with paragraph 13.2(a). In these cases, the fire mains should be located within the protected stairway enclosure, with a maximum hose distance of 45m from the fire main outlet to the furthest point inside each flat, measured on a route suitable for laying a hose.

Design and construction of fire mains

14.4 The outlets from fire mains should be located within the protected stairway enclosure (see Diagram 15.1).

14.5 Guidance on the design and construction of fire mains is given in **BS 9990**.

14.6 Buildings with a storey more than 50m above fire service vehicle access level should be provided with wet fire mains. In all other buildings where fire mains are provided, either wet or dry fire mains are suitable.

14.7 Fire service vehicle access to fire mains should be provided as described in paragraphs 13.5 and 13.6.

Provision of private hydrants

14.8 A building requires additional fire hydrants if both of the following apply.

   a.   It has a compartment with an area of more than 280m^2.

   b.   It is being erected more than 100m from an existing fire hydrant.

14.9 If additional hydrants are required, these should be provided in accordance with the following.

   a.   For buildings provided with fire mains – within 90m of dry fire main inlets.

   b.   For buildings not provided with fire mains – hydrants should be both of the following.

      i.   Within 90m of an entrance to the building.

      ii.   A maximum of 90m apart.

14.10 Each fire hydrant should be clearly indicated by a plate, fixed nearby in a conspicuous position, in accordance with **BS 3251**.

14.11 Guidance on aspects of the provision and siting of private fire hydrants is given in **BS 9990**.

Alternative supply of water

14.12 An alternative source of water should be supplied where any of the following apply.

   a.   No piped water supply is available.

   b.   Pressure and flow in the water main are insufficient.

   c.   An alternative source of supply is proposed.

14.13 The alternative source of water supply should be one of the following, subject to consultation with the local fire and rescue service.

   a.   A charged static water tank with a minimum capacity of 45,000 litres.

   b.   A spring, river, canal or pond that is capable of fulfilling both of the following conditions.

      i.   Providing or storing a minimum of 45,000 litres of water at all times.

      ii.   Providing access, space and a hardstanding for a pumping appliance.

   c.   Any other water supply that the local fire and rescue service considers appropriate.

Section 15: Access to buildings for firefighting personnel – flats

Provision of firefighting shafts

15.1 In low rise buildings without deep basements, access for firefighting personnel is typically achieved by providing measures for fire service vehicle access in Section 13 and means of escape.

15.2 A building with a storey more than 18m above the fire and rescue service vehicle access level should have one or more firefighting shafts, each containing a firefighting lift (Diagram 15.1). The number and location of firefighting shafts should comply with paragraphs 15.4 to 15.7. Firefighting shafts are not required to serve a basement that is not large or deep enough to need one (see paragraph 15.3 and Diagram 15.2).

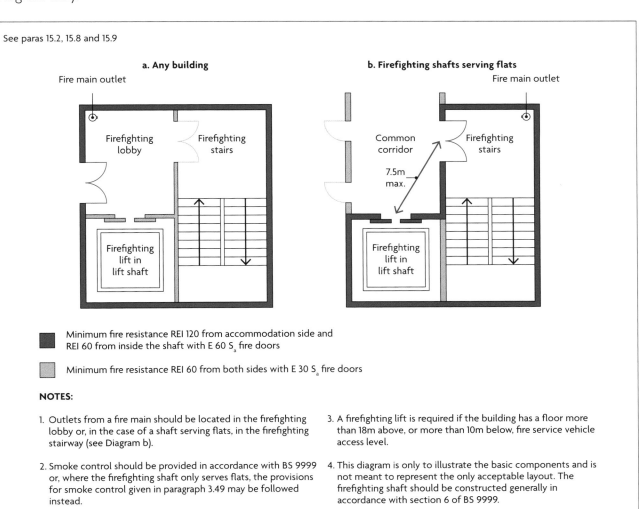

See paras 15.2, 15.8 and 15.9

a. Any building

Fire main outlet

Firefighting lobby

Firefighting stairs

Firefighting lift in lift shaft

b. Firefighting shafts serving flats

Fire main outlet

Common corridor

7.5m max.

Firefighting stairs

Firefighting lift in lift shaft

■ Minimum fire resistance REI 120 from accommodation side and REI 60 from inside the shaft with E 60 S_a fire doors

■ Minimum fire resistance REI 60 from both sides with E 30 S_a fire doors

NOTES:

1. Outlets from a fire main should be located in the firefighting lobby or, in the case of a shaft serving flats, in the firefighting stairway (see Diagram b).

2. Smoke control should be provided in accordance with BS 9999 or, where the firefighting shaft only serves flats, the provisions for smoke control given in paragraph 3.49 may be followed instead.

3. A firefighting lift is required if the building has a floor more than 18m above, or more than 10m below, fire service vehicle access level.

4. This diagram is only to illustrate the basic components and is not meant to represent the only acceptable layout. The firefighting shaft should be constructed generally in accordance with section 6 of BS 9999.

5. For the minimum fire resistance of lift doors see Table C1.

Diagram 15.1 Components of a firefighting shaft

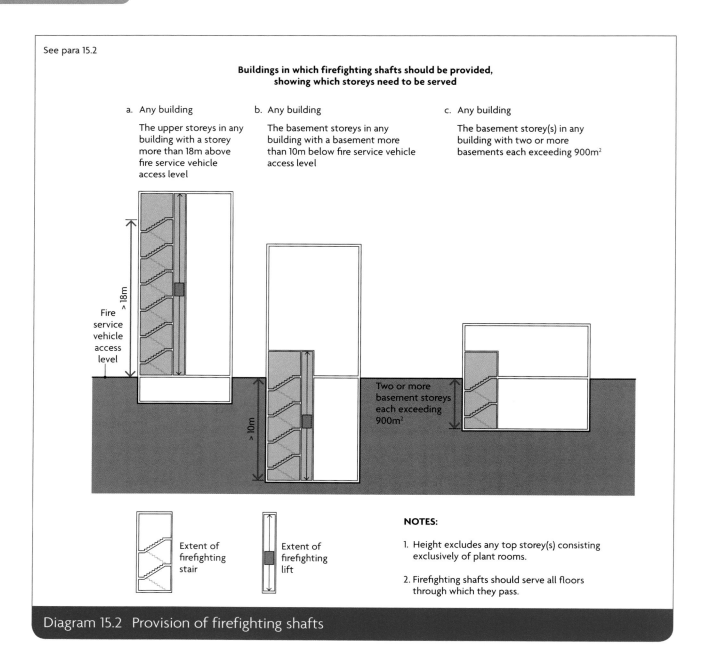

See para 15.2

Buildings in which firefighting shafts should be provided, showing which storeys need to be served

a. Any building

The upper storeys in any building with a storey more than 18m above fire service vehicle access level

b. Any building

The basement storeys in any building with a basement more than 10m below fire service vehicle access level

c. Any building

The basement storey(s) in any building with two or more basements each exceeding 900m²

Fire service vehicle access level

> 18m

> 10m

Two or more basement storeys each exceeding 900m²

Extent of firefighting stair

Extent of firefighting lift

NOTES:

1. Height excludes any top storey(s) consisting exclusively of plant rooms.

2. Firefighting shafts should serve all floors through which they pass.

Diagram 15.2 Provision of firefighting shafts

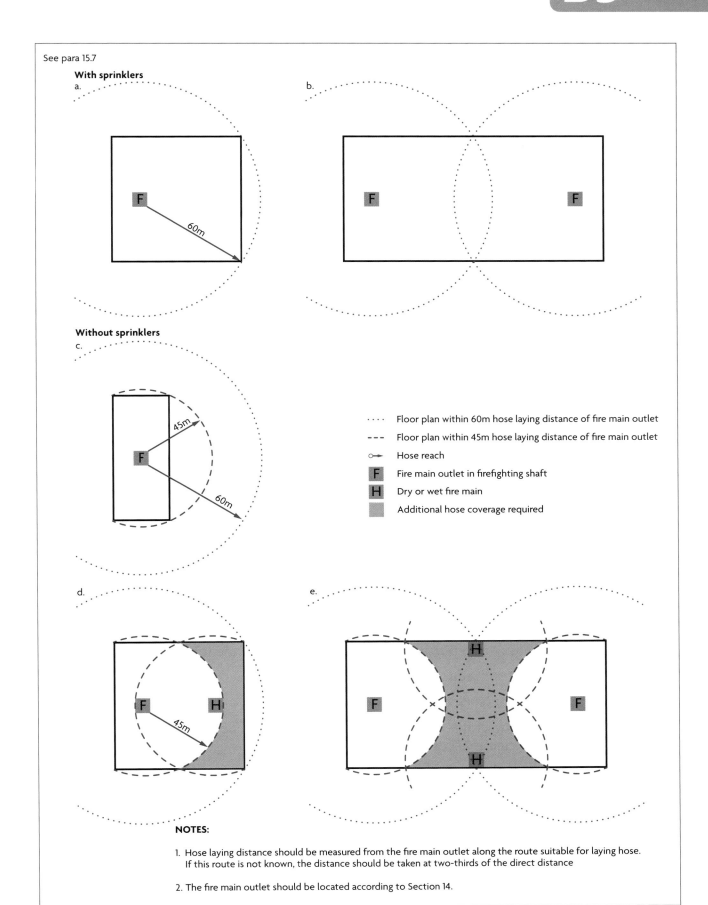

See para 15.7

With sprinklers

a.

b.

Without sprinklers

c.

- - - - Floor plan within 60m hose laying distance of fire main outlet

- - - Floor plan within 45m hose laying distance of fire main outlet

○━ Hose reach

F Fire main outlet in firefighting shaft

H Dry or wet fire main

Additional hose coverage required

d.

e.

NOTES:

1. Hose laying distance should be measured from the fire main outlet along the route suitable for laying hose. If this route is not known, the distance should be taken at two-thirds of the direct distance

2. The fire main outlet should be located according to Section 14.

Diagram 15.3 Location of firefighting shafts: hose laying distances

15.3 A building with basement storeys should have firefighting shafts in accordance with the following.

    a.   There is a basement more than 10m below the fire and rescue service vehicle access level. The firefighting shafts should contain firefighting lifts.

    b.   There are two or more basement storeys, each with a minimum area of 900m². The firefighting shafts do not need to include firefighting lifts.

The building's height and size determine whether firefighting shafts also serve upper storeys.

15.4 Firefighting shafts should serve all storeys through which they pass.

15.5 A minimum of two firefighting shafts should be provided to buildings with a storey that has both of the following.

    a.   A floor area of 900m² or more.

    b.   A floor level 18m or more above the fire and rescue service vehicle access level.

15.6 Firefighting shafts and protected stairways should be positioned such that every part of each storey more than 18m above the fire and rescue service vehicle access level complies with the maximum distances given in paragraph 15.7. Distances should be measured from the fire main outlet on a route suitable for laying a hose.

NOTE: If the internal layout is not known, the distance should be measured at two-thirds of the direct distance.

15.7 In any building, the hose laying distance should meet all of the following conditions.

    a.   A maximum of 60m from the fire main outlet in a firefighting shaft (see Diagram 15.3).

    b.   Additionally, where sprinklers have not been provided in accordance with Appendix E, the hose laying distance should be a maximum of 45m from a fire main outlet in a protected stairway (although this does not imply that the protected stairway needs to be designed as a firefighting shaft (see Diagram 15.3)).

Design and construction of firefighting shafts

15.8 Firefighting stairs and firefighting lifts should be approached from either of the following.

    a.   A firefighting lobby.

    b.   A protected corridor or protected lobby that complies with the following guidance.

        i.   Means of escape (Section 3).

        ii.   Compartmentation (Section 7).

Both the stair and lobby of the firefighting shaft should be provided with a means of venting smoke and heat (see clause 27.1 of **BS 9999**).

Only services associated with the firefighting shaft, such as ventilation systems and lighting for the firefighting shafts, should pass through or be contained within the firefighting shaft.

Doors of a firefighting lift landing should be a maximum of 7.5m from the door to the firefighting stair (Diagram 15.1).

15.9 Firefighting shafts should achieve a minimum fire resistance of REI 120. A minimum of REI 60 is acceptable for either of the following (see Diagram 15.1).

    a. Constructions separating the firefighting shaft from the rest of the building.

    b. Constructions separating the firefighting stair, firefighting lift shaft and firefighting lobby.

15.10 All firefighting shafts should have fire mains with outlet connections and valves at every storey.

15.11 A firefighting lift installation includes all of the following.

    a. Lift car.

    b. Lift well.

    c. Lift machinery space.

    d. Lift control system.

    e. Lift communications system.

The lift shaft should be constructed in accordance with Section 6 of **BS 9999**.

Firefighting lift installations should conform to **BS EN 81-72** and **BS EN 81-20**.

Rolling shutters in compartment walls

15.12 The fire and rescue service should be able to manually open and close rolling shutters without the use of a ladder.

B5

Section 16: Venting of heat and smoke from basements – flats

Provision of smoke outlets

16.1 Heat and smoke from basement fires vented via stairs can inhibit access for firefighting personnel. This may be reduced by providing smoke outlets, or smoke vents, which allow heat and smoke to escape from the basement levels to the open air. They can also be used by the fire and rescue service to let cooler air into the basements (Diagram 16.1).

16.2 Each basement space should have one or more smoke outlets.

Where this is not practicable (for example, the plan area is deep and the amount of external wall is restricted by adjoining buildings), the perimeter basement spaces may be vented, with other spaces vented indirectly by opening connecting doors. This does not apply for places of special fire hazard (see paragraph 16.7).

If a basement is compartmented, each compartment should have one or more smoke outlets, rather than indirect venting.

A basement storey or compartment containing rooms with doors or windows does not need smoke outlets.

16.3 Smoke outlets connecting directly to the open air should be provided from every basement storey, except for any basement storey that has both of the following.

a. A maximum floor area of 200m².

b. A floor a maximum of 3m below the adjacent ground level.

16.4 Strong rooms do not need to be provided with smoke outlets.

Natural smoke outlets

16.5 Smoke outlets should be both of the following.

a. Sited at high level in either the ceiling or wall of the space they serve.

b. Evenly distributed around the perimeter, to discharge to the open air.

16.6 The combined clear cross-sectional area of all smoke outlets should be a minimum of 1/40 of the area of the floor of the storey they serve.

16.7 Separate outlets should be provided from places of special fire hazard.

16.8 If the smoke outlet terminates at a point that is not readily accessible, it should be kept unobstructed and covered only with a class A1 grille or louvre.

16.9 If the smoke outlet terminates in a readily accessible position, it may be covered by a panel, stallboard or pavement light that can be broken out or opened. The position of covered smoke outlets should be suitably indicated.

16.10 Outlets should not be placed where they prevent the use of escape routes from the building.

Mechanical smoke extract

16.11 If basement storeys are fitted with a sprinkler system in accordance with Appendix E, a mechanical smoke extraction system may be provided as an alternative to natural venting. Sprinklers do not need to be installed on the other storeys unless needed for other reasons.

Car parks are not normally expected to be fitted with sprinklers (see Section 11 of Approved Document B Volume 2).

16.12 The air extraction system should comply with all of the following.

a. It should give at least 10 air changes per hour.

b. It should be capable of handling gas temperatures of 300°C for not less than one hour.

c. It should do either of the following.

    i. Be activated automatically if the sprinkler system activates.

    ii. Be activated by an automatic fire detection system that conforms to **BS 5839-1** (minimum L3 standard).

Further information on equipment for removing hot smoke is given in **BS EN 12101-3**.

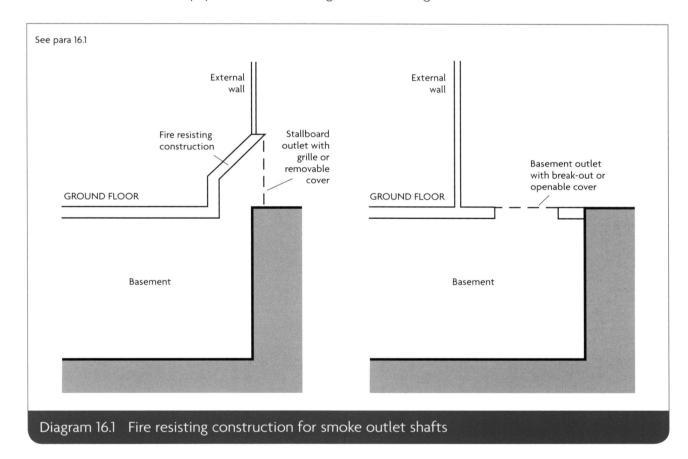

See para 16.1

External wall

Fire resisting construction

Stallboard outlet with grille or removable cover

GROUND FLOOR

Basement

External wall

Basement outlet with break-out or openable cover

GROUND FLOOR

Basement

Diagram 16.1   Fire resisting construction for smoke outlet shafts

Construction of outlet ducts or shafts

16.13 Outlet ducts or shafts, including any bulkheads over them (see Diagram 16.1), should be enclosed in construction of class A1 rating and fire resistance at least equal to that of the element through which they pass.

16.14 Natural smoke outlet shafts should be separated from each other using construction of class A1 rating and fire resistance at least equal to that of the storeys they serve, where the shafts are either of the following.

a. From different compartments of the same basement storey.

b. From different basement storeys.

Regulation 38: Fire safety information

This section deals with the following regulation of the Building Regulations 2010.

Fire safety information

38. (1) This regulation applies where building work—

    (a) consists of or includes the erection or extension of a relevant building; or

    (b) is carried out in connection with a relevant change of use of a building,

    and Part B of Schedule 1 imposes a requirement in relation to the work.

(2) The person carrying out the work shall give fire safety information to the responsible person not later than the date of completion of the work, or the date of occupation of the building or extension, whichever is the earlier.

(3) In this regulation—

    (a) "fire safety information" means information relating to the design and construction of the building or extension, and the services, fittings and equipment provided in or in connection with the building or extension which will assist the responsible person to operate and maintain the building or extension with reasonable safety;

    (b) a "relevant building" is a building to which the Regulatory Reform (Fire Safety) Order 2005 applies, or will apply after the completion of building work;

    (c) a "relevant change of use" is a material change of use where, after the change of use takes place, the Regulatory Reform (Fire Safety) Order 2005 will apply, or continue to apply, to the building; and

    (d) "responsible person" has the meaning given by article 3 of the Regulatory Reform (Fire Safety) Order 2005.

Intention

The aim of this regulation is to ensure that the person responsible for the building has sufficient information relating to fire safety to enable them to manage the building effectively. The aim of regulation 38 will be achieved when the person responsible for the building has all the information to enable them to do all of the following.

a. Understand and implement the fire safety strategy of the building.

b. Maintain any fire safety system provided in the building.

c. Carry out an effective fire risk assessment of the building.

Section 17: Fire safety information

17.1 For building work involving the erection or extension of a relevant building (i.e. a building to which the Regulatory Reform (Fire Safety) Order 2005 applies, or will apply), or the relevant change of use of a building, fire safety information should be given to the responsible person at one of the following times.

    a. When the project is complete.

    b. When the building or extension is first occupied.

17.2 This section is a guide to the information that should be provided. Guidance is in terms of essential information and additional information for complex buildings; however, the level of detail required should be considered on a case-by-case basis.

Essential information

17.3 Basic information on the location of fire protection measures may be sufficient. An as-built plan of the building should be provided showing all of the following.

    a. Escape routes – this should include exit capacity (i.e. the maximum allowable number of people for each storey and for the building).

    b. Location of fire-separating elements (including cavity barriers in walk-in spaces).

    c. Fire doorsets, fire doorsets fitted with a self-closing device and other doors equipped with relevant hardware.

    d. Locations of fire and/or smoke detector heads, alarm call points, detection/alarm control boxes, alarm sounders, fire safety signage, emergency lighting, fire extinguishers, dry or wet fire mains and other firefighting equipment, and hydrants outside the building.

    e. Any sprinkler systems, including isolating valves and control equipment.

    f. Any smoke control systems, or ventilation systems with a smoke control function, including mode of operation and control systems.

    g. Any high risk areas (e.g. heating machinery).

17.4 Details should be provided of all of the following.

    a. Specifications of fire safety equipment provided, including routine maintenance schedules.

    b. Any assumptions regarding the management of the building in the design of the fire safety arrangements.

    c. Any provision enabling the evacuation of disabled people, which can be used when designing personal emergency evacuation plans.

Additional information for complex buildings

17.5 A detailed record should be provided of both of the following.

    a. The fire safety strategy.

    b. Procedures for operating and maintaining any fire protection measures. This should include an outline cause and effect matrix/strategy for the building.

Further guidance is available in clause 9 and Annex H of **BS 9999**.

17.6 The records should include details of all of the following.

    a. The fire safety strategy, including all assumptions in the design of the fire safety systems (such as fire load). Any risk assessments or risk analysis.

    b. All assumptions in the design of the fire safety arrangements for the management of the building.

    c. All of the following.

        i. Escape routes (including occupant load and capacity of escape routes).

        ii. Any provision to enable the evacuation of disabled people.

        iii. Escape strategy (e.g. simultaneous or phased).

        iv. Muster points.

    d. All passive fire safety measures, including all of the following.

        i. Compartmentation (i.e. location of fire-separating elements).

        ii. Cavity barriers.

        iii. Fire doorsets, including fire doorsets fitted with a self-closing device and other doors equipped with relevant hardware (e.g. electronic security locks).

        iv. Duct dampers.

        v. Fire shutters.

    e. All of the following.

        i. Fire detector heads.

        ii. Smoke detector heads.

        iii. Alarm call points.

        iv. Detection/alarm control boxes.

        v. Alarm sounders.

        vi. Emergency communications systems.

        vii. CCTV.

        viii. Fire safety signage.

        ix. Emergency lighting.

        x. Fire extinguishers.

        xi. Dry or wet fire mains and other firefighting equipment.

    xii.   Other interior facilities for the fire and rescue service.

    xiii. Emergency control rooms.

    xiv. Location of hydrants outside the building.

    xv. Other exterior facilities for the fire and rescue service.

f. All active fire safety measures, including both of the following.

    i. Sprinkler system(s) design, including isolating valves and control equipment.

    ii. Smoke control system(s) (or heating, ventilation and air conditioning system with a smoke control function) design, including mode of operation and control systems.

g. Any high risk areas (e.g. heating machinery) and particular hazards.

h. Plans of the building as built, showing the locations of the above.

i. Both of the following.

    i. Specifications of any fire safety equipment provided, including all of the following.

- Operational details.
- Operators' manuals.
- Software.
- System zoning.
- Routine inspection, testing and maintenance schedules.

    ii. Records of any acceptance or commissioning tests.

j. Any other details appropriate for the specific building.

Appendix A: Key terms

NOTE: Except for the items marked * (which are from the Building Regulations 2010), these definitions apply only to Approved Document B.

NOTE: The terms defined below are key terms used in this document only. Refer to **BS 4422** for further guidance on the definitions of common terms used in the fire safety industry which are not listed below.

Access room A room that the only escape route from an inner room passes through.

Alternative escape routes Escape routes that are sufficiently separated by direction and space or by fire resisting construction to ensure that one is still available if the other is affected by fire.

NOTE: A second stair, balcony or flat roof which enables a person to reach a place free from danger from fire is considered an alternative escape route for the purposes of a dwellinghouse.

Alternative exit One of two or more exits, each of which is separate from the other.

Appliance ventilation duct A duct to deliver combustion air to a gas appliance.

Atrium (plural **atria**) A continuous space that passes through one or more structural floors within a building, not necessarily vertically.

NOTE: Enclosed lift wells, enclosed escalator wells, building services ducts and stairs are not classified as atria.

Automatic release mechanism A device that normally holds a door open, but closes it automatically if any one of the following occurs.

- Smoke is detected by an automatic device of a suitable nature and quality in a suitable location.

- A hand-operated switch, fitted in a suitable position, is operated.

- The electricity supply to the device, apparatus or switch fails.

- The fire alarm system, if any, is operated.

Basement storey A storey with a floor that, at some point, is more than 1200mm below the highest level of ground beside the outside walls. (However, see Appendix B, paragraph B26c, for situations where the storey is considered to be a basement only because of a sloping site.)

Boundary The boundary of the land that belongs to a building, or, where the land abuts a road, railway, canal or river, the centre line of that road, railway, canal or river.

*Building Any permanent or temporary building but not any other kind of structure or erection. A reference to a building includes a reference to part of a building.

Building control body A term that includes both local authority building control and approved inspectors.

Cavity A space enclosed by elements of a building (including a suspended ceiling) or contained within an element, but that *is not* a room, cupboard, circulation space, protected shaft, or space within a flue, chute, duct, pipe or conduit.

Cavity barrier A construction within a cavity, other than a smoke curtain, to perform either of the following functions.

- Close a cavity to stop smoke or flame entering.

- Restrict the movement of smoke or flame within a cavity.

Ceiling Part of a building that encloses a room, protected shaft or circulation space and is exposed overhead.

NOTE: The soffit of a rooflight, but not the frame, is included as part of the surface of the ceiling. An upstand below a rooflight is considered as a wall.

Circulation space A space (including a protected stairway) mainly used as a means of access between a room and an exit from the building or compartment.

Common balcony A walkway, open to the air on one or more sides, that forms part of the escape route from more than one flat.

Common stair An escape stair that serves more than one flat.

Compartment (fire) A building or part of a building, comprising one or more rooms, spaces or storeys, that is constructed to prevent the spread of fire to or from another part of the same building or an adjoining building.

NOTE: A roof space above the top storey of a compartment is included in that compartment. (See also 'Separated part'.)

Compartment wall or floor A fire resisting wall or floor to separate one fire compartment from another.

NOTE: Provisions relating to construction are given in Section 7.

Corridor access A design of a building containing flats, in which each flat is approached via a common horizontal internal access or circulation space, which may include a common entrance hall.

Dead end An area from which escape is possible in one direction only.

Direct distance The shortest distance from any point within the floor area to the nearest storey exit, measured within the external enclosures of the building, and ignoring walls, partitions and fittings other than the enclosing walls and partitions to protected stairways.

*Dwelling Includes a dwellinghouse and a flat.

NOTE: A dwelling is a unit where one or more people live (whether or not as a sole or main residence) in either of the following situations.

- A single person or people living together as a family.

- A maximum of six people living together as a single household, including where care is provided for residents.

*Dwellinghouse Does not include a flat or a building containing a flat.

Element of structure Any of the following.

- A member that forms part of the structural frame of a building, or any other beam or column.

- A loadbearing wall or loadbearing part of a wall.

- A floor.

- A gallery (but *not* a loading gallery, fly gallery, stage grid, lighting bridge, or any gallery provided for similar purposes or for maintenance and repair).

- An external wall.

- A compartment wall (including a wall that is common to two or more buildings).

NOTE: However, see the guidance to requirement B3, paragraph 6.2, for a list of structures that are *not* considered to be elements of structure.

Emergency lighting Lighting for use when the power supply to the normal lighting fails.

Escape lighting The part of the emergency lighting that is provided to ensure that the escape route is illuminated at all material times.

Escape route The route along which people can escape from any point in a building to a final exit.

Evacuation lift A lift that may be used to evacuate people in a fire.

Exit passageway A protected passageway that connects a protected stairway to a final exit.

NOTE: Exit passageways should be protected to the same standard as the stairway they serve.

*External wall The external wall of a building includes all of the following.

- Anything located within any space forming part of the wall.

- Any decoration or other finish applied to any external (but not internal) surface forming part of the wall.

- Any windows and doors in the wall.

- Any part of a roof pitched at an angle of more than 70 degrees to the horizontal if that part of the roof adjoins a space within the building to which persons have access, but not access only for the purpose of carrying out repairs or maintenance.

Final exit The end of an escape route from a building that gives direct access to a street, passageway, walkway or open space, and is sited to ensure that people rapidly disperse away from the building so that they are no longer in danger from fire and/or smoke.

NOTE: Windows are not acceptable as final exits.

Fire alarm system Combination of components for giving an audible and/or other perceptible warning of fire.

Fire damper A mechanical or intumescent device within a duct or ventilation opening that operates automatically and is designed to resist the spread of fire.

Fire and smoke damper A fire damper which, in addition to the performance of the fire damper, resists the spread of smoke.

Fire doorset A door or shutter which, together with its frame and furniture as installed in a building, is intended (when closed) to resist the spread of fire and/or gaseous products of combustion and meets specified performance criteria to those ends.

NOTE: A fire doorset may have one or more leaves. The term includes a cover or other form of protection to an opening in a fire resisting wall or floor, or in a structure that surrounds a protected shaft. A fire doorset is a complete door assembly, assembled on site or delivered as a completed assembly, consisting of the door frame, leaf or leaves, essential hardware, edge seals and glazing, and any integral side panels or fanlight panels in an associated door screen.

Firefighting lift A lift with additional protection and with controls that enable it to be used by the fire and rescue service when fighting a fire. (See Section 15.)

Firefighting lobby A protected lobby that provides access from a firefighting stair to the accommodation area and to any associated firefighting lift.

Firefighting shaft A protected enclosure that contains a firefighting stair, firefighting lobbies and, if provided, a firefighting lift together with its machine room.

Firefighting stair A protected stairway that connects to the accommodation area through only a firefighting lobby.

Fire resisting (Fire resistance) The ability of a component or a building to satisfy, for a stated period of time, some or all of the appropriate criteria given in the relevant standard.

Fire-separating element A compartment wall, compartment floor, cavity barrier and construction that encloses a protected escape route and/or a place of special fire hazard.

Fire-stop (Fire-stopping) A seal provided to close an imperfection of fit or design tolerance between elements or components, to restrict the spread of fire and smoke.

*Flat A flat is a separate and self-contained premises constructed or adapted for use for residential purposes and forming part of a building from some other part of which it is divided horizontally.

Gallery A floor or balcony that does not extend across the full extent of a building's footprint and is open to the floor below.

Habitable room A room used, or intended to be used, for people to live in (including, for the purposes of Approved Document B Volumes 1 and 2, a kitchen, but not a bathroom).

Height (of a building or storey for the purposes of Approved Document B Volumes 1 and 2)

- Height of a building is measured as shown in Appendix D, Diagram D4.
- Height of the floor of the top storey above ground level is measured as shown in Appendix D, Diagram D6.

Inner room Room from which escape is possible only by passing through another room (the access room).

Live/work unit A flat that is a workplace for people who live there, its occupants, and for people who do not live on the premises.

Means of escape Structural means that provide one or more safe routes for people to go, during a fire, from any point in the building to a place of safety.

Measurement

- Width of a doorway, cubic capacity, area, height of a building and number of storeys are measured as shown in Appendix D, Diagrams D1 to D6.

- Occupant number, travel distance, escape route and stairs are measured as described in Appendix D, paragraphs D1 to D4.

Notional boundary A boundary presumed to exist between two buildings on the same site.

Open spatial planning The internal arrangement of a building in which more than one storey or level is contained in one undivided volume, e.g. split-level floors. For the purposes of this document there is a distinction between open spatial planning and an atrium space.

Perimeter (of a building) The maximum aggregate plan perimeter, found by vertical projection onto a horizontal plane. (See Section 15 of Approved Document B Volume 2.)

Pipe Includes pipe fittings and accessories. The definition of 'pipe' *excludes* a flue pipe and a pipe used for ventilating purposes, other than a ventilating pipe for an above-ground drainage system.

Place of special fire hazard A room such as any of the following.

- Oil-filled transformer room.

- Switch gear room.

- Boiler room.

- Storage space for fuel or other highly flammable substance(s).

- Room that houses a fixed internal combustion engine.

Platform floor (also called an access or raised floor) A floor that is supported by a structural floor, but with an intervening cavity to house services.

Protected circuit An electrical circuit that is protected against fire.

Protected corridor/lobby A corridor or lobby that is adequately protected from fire in adjoining areas by fire resisting construction.

Protected entrance hall/landing A circulation area, consisting of a hall or space in a flat, that is enclosed with fire resisting construction other than an external wall of a building.

Protected shaft A shaft that enables people, air or objects to pass from one compartment to another, and which is enclosed with fire resisting construction.

Protected stairway A stair that leads to a final exit to a place of safety and that is adequately enclosed with fire resisting construction. Included in the definition is any exit passageway between the foot of the stair and the final exit.

Purpose group A classification of a building according to the purpose to which it is intended to be put. (See Table 0.1.)

Relevant boundary The boundary or notional boundary that one side of the building faces and/or coincides with, and that is parallel or at an angle of a maximum of 80 degrees to that side of the building.

Rooflight A dome light, lantern light, skylight, ridge light, glazed barrel vault or other element to admit daylight through a roof.

Room An enclosed space within a building that is not used solely as a circulation space. The term includes not only conventional rooms, but also cupboards that are not fittings and large spaces such as warehouses and auditoria. The term *does not* include cavities such as ducts, ceiling cavities and roof spaces.

School A place of education for children between 2 and 19 years old. The term includes nursery schools, primary schools and secondary schools as defined in the Education Act 1996.

Self-closing device A device that closes a door, when open at any angle, against a door frame.

NOTE: If the door is in a cavity barrier, rising butt hinges (which are different from the self-closing device mentioned above) are acceptable.

Separated part (of a building) Part of a building that is separated from another part of the same building by a compartment wall. The wall runs the full height of the part and is in one vertical plane. (See Appendix D, Diagram D5.)

Sheltered housing Includes two or more dwellings in the same building or on adjacent sites, designed and constructed as residential accommodation for vulnerable or elderly people who receive, or will receive, a support service.

Single storey building A building that consists of a ground storey only. Basements are not counted as storeys in a building (see Appendix D). A separated part that consists of a ground storey only, with a roof to which access is only provided for repair or maintenance, may be treated as a single storey building.

Site (of a building) The land occupied by the building, up to the boundaries with land in other ownership.

*Specified attachment Includes any of the following.

- A balcony attached to an external wall.

- A device for reducing heat gain within a building by deflecting sunlight which is attached to an external wall.

- A solar panel attached to an external wall.

Storey Includes any of the following.

- Any gallery in an assembly building (purpose group 5).

- Any gallery in any other type of building if its area is more than half that of the space into which it projects.

- A roof, unless it is accessible only for maintenance and repair.

NOTE: The building is regarded as a multi-storey building if both of the following apply.

- There is more than one gallery.

- The total aggregate area of all the galleries in one space is more than half the floor area of that space.

Storey exit A final exit, or a doorway that gives direct access into a protected stairway, firefighting lobby or external escape route.

NOTE: If an institutional building is planned to enable progressive horizontal evacuation, a door in a compartment wall is considered a storey exit for the purposes of requirement B1.

Suspended ceiling (fire-protecting) A ceiling suspended below a floor that adds to the fire resistance of the floor.

Thermoplastic material Any synthetic polymeric material that has a softening point below 200°C if tested to **BS EN ISO 306** Method A120. Specimens for this test may be fabricated from the original polymer where the thickness of material of the end product is less than 2.5mm.

Travel distance (unless otherwise specified, e.g. as in the case of flats) The distance that a person would travel from any point within the floor area to the nearest storey exit, determined by the layout of walls, partitions and fittings.

Unprotected area (in relation to a side or external wall of a building) All of the following are classed as unprotected areas.

- Any part of the external wall that has less than the relevant fire resistance set out in Section 11.

- Any part of the external wall constructed of material more than 1mm thick if that material does not have a class B-s3, d2 rating or better, which is attached or applied, whether for cladding or any other purpose.

- Windows, doors or other openings. This does not include windows that are designed and glazed to give the necessary level of fire resistance and that are not openable.

NOTE: Recessed car parking areas as shown in Diagram A1 should not be regarded as unprotected areas.

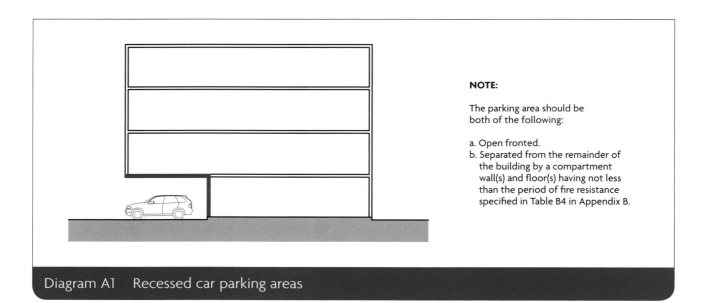

NOTE:

The parking area should be both of the following:

a. Open fronted.
b. Separated from the remainder of the building by a compartment wall(s) and floor(s) having not less than the period of fire resistance specified in Table B4 in Appendix B.

Diagram A1   Recessed car parking areas

Appendix B: Performance of materials, products and structures

Introduction

B1   Much of the guidance in this document is given in terms of performance classifications in relation to British or European Standards. In such cases, it will be necessary to demonstrate that a system or product can meet the relevant performance classification. This will be achieved if the system or product complies with one of the following.

    a.   They should be in accordance with a specification or design that has been shown by a specific test to be capable of meeting that performance classification.

    b.   They should have been designed by using relevant design standards in order to meet that performance classification.

    c.   They should have been assessed by applying relevant test evidence, in lieu of carrying out a specific test, as being capable of meeting that performance classification.

    NOTE: Some products are subject to Classification Without Further Testing (CWFT). For the purposes of this approved document, such products can be considered to have been shown to be capable of meeting a performance specification as per paragraph B1a.

B2   Any test evidence used to demonstrate the fire performance classification of a product or system should be carefully checked to ensure that it is applicable to the intended use. Small differences in detail, such as fixing method, joints, dimensions, the introduction of insulation materials and air gaps (ventilated or not), can significantly affect the performance.

B3   Assessments should not be regarded as a way to avoid a test where one is necessary. Assessments should only be carried out where sufficient relevant test evidence is available. Relevant test evidence is unlikely to be provided by test standards which have different classification criteria.

B4   Where it is proposed to assess the classification of a product or system in lieu of carrying out a specific test (as in paragraph B1b), this should be done in accordance with the relevant standard for extended application for the test in question and should include details of the test evidence that has been used to support the assessment.

For performance classifications where there is no specific standard for extended application, assessment reports should be produced in accordance with the principles of **BS EN 15725** and should include details of the test evidence that has been used to support the assessment. Further information on best practice is provided in the Passive Fire Protection Federation's *Guide to Undertaking Assessments in Lieu of Fire Tests*.

    NOTE: Regulation 7(2) limits components used in or on the external walls of certain buildings to materials achieving class A2-s1, d0 or class A1 (see Section 10). Assessments cannot be used to demonstrate compliance with this requirement.

B5   Tests and assessments should be carried out by organisations with the necessary expertise. For example, organisations listed as 'notified bodies' in accordance with the European Construction

Products Regulation or laboratories accredited by the United Kingdom Accreditation Service (UKAS) for the relevant test standard can be assumed to have the necessary expertise.

NOTE: Standard fire tests do not directly measure fire hazard. They measure or assess the response of a material or system to exposure to one or more aspects of fire conditions. Performance in fire tests is only one of a number of factors that should be taken into account.

Reaction to fire

B6   Reaction to fire relates to the degree to which a product will contribute, by its own decomposition, to a fire under specified conditions. Products, other than floorings, are classified as A1, A2, B, C, D, E or F (with class A1 being the highest performance and F being the lowest) in accordance with **BS EN 13501-1**. Class F is assigned when a product fails to attain class E. Untested products cannot be classified in accordance with **BS EN 13501-1**.

Materials covered by the Classification Without Further Testing (CWFT) process can be found by accessing the European Commission's website https://eur-lex.europa.eu/.

B7   The classes of reaction to fire performance of A2, B, C, D and E are accompanied by additional classifications related to the production of smoke (s1, s2, s3), with s1 indicating the lowest production, and/or flaming droplets/particles (d0, d1, d2), with d0 indicating the lowest production.

NOTE: When a classification includes s3, d2 this means that there is no limit set for smoke production and/or flaming droplets/particles.

B8   To reduce the testing burden on manufacturers, **BS EN 13238** defines a number of standard substrates that produce test results representative of different end use applications. The classification for reaction to fire achieved during testing is only valid when the product is used within this field of application, i.e. when the product is fixed to a substrate of that class in its end use. The standard substrate selected for testing should take account of the intended end use applications (field of application) of the product and represent end use substrates that have a density of a minimum of 75% of the standard substrate's nominal density.

B9   Standard substrates include gypsum plasterboard (**BS EN 520**) with a density of 700+/-100kg/m^3, calcium silicate board (**BS EN 14306**) 870+/-50kg/m^3 and fibre-cement board 1800+/-200kg/m^3.

NOTE: Standard calcium silicate board is not representative of gypsum plasterboard end use (due to the paper layer), but would be representative of most gypsum plasters (with densities of more than 650kg/m^3).

NOTE: Classifications based on tests using a plasterboard substrate would also be acceptable for products bonded to a gypsum plaster end use substrate.

National classifications for reaction to fire

B10 This document uses the European classification system for reaction to fire set out in **BS EN 13501-1**; however, there may be some products lawfully on the market using the classification system set out in previous editions. Where this is the case, Table B1 can be used for the purposes of this document.

Table B1 Reaction to fire classifications: transposition to national class

BS EN 13501-1 classification	Transposition
A1	Material that, when tested to **BS 476-11**, does not either: a. flame b. cause a rise in temperature on either the thermocouple at the centre of the specimen or in the furnaces
A2-s1, d0	None

A2-s3, d2	Material that meets either of the following.
	
	a. Any material of density 300kg/m³ or more, which, when tested to **BS 476-11**, complies with both of the following:
	
	  i. does not flame
	
	  ii. causes a rise in temperature on the furnace thermocouple not exceeding 20°C
	
	b. Any material of density less than 300kg/m³, which, when tested to **BS 476-11**, complies with both of the following:
	
	  i. does not flame for more than 10 seconds
	
	  ii. causes a rise in temperature on the thermocouple at the centre of the specimen or in the furnace that is a maximum of 35°C and on the furnace thermocouple that is a maximum of 25°C

B-s3, d2	Any material that meets both of the following criteria. a. Class 1 in accordance with **BS 476-7**. b. Has a fire propagation index (I) of a maximum of 12 and sub-index (i1) of a maximum of 6, determined by using the method given in **BS 476-6**. Index of performance (I) relates to the overall test performance, whereas sub-index (i1) is derived from the first three minutes of the test
C-s3, d2	Class 1 in accordance with **BS 476-7**
D-s3, d2	Class 3 in accordance with **BS 476-7**

NOTE: The national classifications do not automatically equate with the transposed classifications in the '**BS EN 13501-1** classification' column, therefore products cannot typically assume a European class unless they have been tested accordingly.

NOTE: A classification of s3, d2 indicates that no limit is set for production of smoke and/or flaming droplets/particles. If a performance for production of smoke and/or flaming droplets/particles is specified, then only the European classes can be used. For example, a national class may not be used as an alternative to a classification which includes s1, d0.

Thermoplastic materials

B11 Thermoplastic material is any synthetic polymeric material that has a softening point below 200°C if tested to **BS EN ISO 306** Method A120. Products formed from these materials cannot always be classified in the normal way. In those circumstances the following approach can be followed.

B12 Thermoplastic materials used for window glazing, rooflights and lighting diffusers within suspended ceilings do not need to meet the criteria within paragraph B19 onwards, if the guidance to requirements B2 and B4 is followed.

B13 For the purposes of requirements B2 and B4, thermoplastic materials should be classified as TP(a) rigid, TP(a) flexible or TP(b), as follows:

  a. **TP(a) rigid**

    i. rigid solid uPVC sheet

    ii. solid (as distinct from double- or multi-skinned) polycarbonate sheet a minimum of 3mm thick

    iii. multi-skinned rigid sheet made from uPVC or polycarbonate that has a class 1 rating when tested to **BS 476-7**

    iv. any other rigid thermoplastic product, a specimen of which (at the thickness of the product as put on the market), when tested to **BS 2782-0** Method 508A, performs so that both:

      • the test flame extinguishes before the first mark

      • the duration of flaming or afterglow does not exceed 5 seconds following removal of the burner.

  b. **TP(a) flexible**

    Flexible products a maximum of 1mm thick that comply with the Type C requirements of **BS 5867-2** when tested to **BS 5438** Test 2 with the flame applied to the surface of the specimens for 5, 15, 20 and 30 seconds respectively, but excluding the cleansing procedure; and

  c. **TP(b)**

    i. rigid solid polycarbonate sheet products a maximum of 3mm thick, or multi-skinned polycarbonate sheet products that do not qualify as TP(a) by test

    ii. other products which, when a specimen of the material between 1.5 and 3mm thick is tested in accordance with **BS 2782-0** Method 508A, have a maximum rate of burning of 50mm/minute.

NOTE: If it is not possible to cut or machine a 3mm thick specimen from the product, then a 3mm test specimen can be moulded from the same material as that used to manufacture the product.

B14 A thermoplastic material alone when used as a lining to a wall or ceiling cannot be assumed to protect a substrate. The surface rating of both thermoplastic material and substrate must therefore meet the required classification.

If, however, the thermoplastic material is fully bonded to a non-thermoplastic substrate, then only the surface rating of the composite needs to meet the required classification.

Roofs

B15 Performance of the resistance of roofs to external fire exposure is measured in terms of penetration through the roof construction and the spread of flame over its surface.

B16 Roof constructions are classified within the European system as $B_{ROOF}(t4)$, $C_{ROOF}(t4)$, $D_{ROOF}(t4)$, $E_{ROOF}(t4)$ or $F_{ROOF}(t4)$ in accordance with **BS EN 13501-5**. $B_{ROOF}(t4)$ indicates the highest performance and $F_{ROOF}(t4)$ the lowest.

B17 **BS EN 13501-5** refers to four separate roof tests. The suffix (t4) used in paragraph B16 indicates that Test 4 is to be used for the purposes of this approved document.

B18 This document uses the European classification system for roof covering set out in **BS EN 13501-5**; however, there may be some products lawfully on the market using the classification system set out in previous editions. Where this is the case, Table B2 can be used for the purposes of this document.

Table B2 Roof covering classifications: transposition to national class	
BS EN 13501-5 classification	Transposition to **BS 476-3** classification
B_{ROOF} (t4)	AA, AB or AC
C_{ROOF}(t4)	BA, BB or BC
D_{ROOF}(t4)	CA, CB or CC
E_{ROOF}(t4)	AD, BD or CD
F_{ROOF}(t4)	DA, DB, DC or DD

NOTE: The national classifications do not automatically equate with the transposed classifications in the European column, therefore products cannot typically assume a European class unless they have been tested accordingly.

Fire resistance

B19 Common to all of the provisions of Part B of the Building Regulations is the property of fire resistance. Fire resistance is a measure of one or more of the following.

a. **Resistance to collapse** (loadbearing capacity), which applies to loadbearing elements only, denoted R in the European classification of the resistance to fire performance.

b. **Resistance to fire penetration** (integrity), denoted E in the European classification of the resistance to fire performance.

c. **Resistance to the transfer of excessive heat** (insulation), denoted I in the European classification of the resistance to fire performance.

B20 The standards of fire resistance necessary for a particular building are based on assumptions about the severity of fires and the consequences should an element fail. Fire severity is estimated in very broad terms from the use of the building (its purpose group), on the assumption that the building contents (which constitute the fire load) are similar for buildings with the same use.

B21 Because the use of buildings may change, a precise estimate of fire severity based on the fire load due to a particular use may be misleading. Therefore if a fire engineering approach of this kind is adopted, the likelihood that the fire load may change in the future needs to be considered.

B22 Performance in terms of the fire resistance to be achieved by elements of structure, doors and other forms of construction is classified in accordance with one of the following.

a. **BS EN 13501-2**.

b. **BS EN 13501-3**.

c. **BS EN 13501-4**.

B23 Fire resistance is measured in minutes. This relates to time elapsed in a standard test and should not be confused with real time.

B24 The fire resistance necessary for different circumstances is set out in the following tables.

a. Table B3 gives the specific requirements for each element of structure.

b. Table B4 sets out the minimum periods of fire resistance for elements of structure.

c. Table B5 sets out limitations on the use of uninsulated fire resisting glazed elements.

B25 This document uses the European classification system for fire resistance set out in **BS EN 13501-2 to 4**; however, there may be some products lawfully on the market using the classification system set out in previous editions. In those situations the equivalent classifications given in Table B1 can be used.

Table B3 Specific provisions of the test for fire resistance of elements of structure, etc.

Part of building	Minimum provisions when tested to the relevant European standard (minutes)[1]	Alternative minimum provisions when tested to the relevant part of **BS 476**[2] (minutes)			Type of exposure
		Loadbearing capacity[3]	Integrity	Insulation	
1. **Structural** frame, beam or column.	R see Table B4	See Table B4	Not applicable	Not applicable	Exposed faces
2. **Loadbearing wall** (which is not also a wall described in any of the following items).	R see Table B4	See Table B4	Not applicable	Not applicable	Each side separately
3. **Floors**[4]					
a. between a shop and flat above	REI 60 or see Table B4 (whichever is greater)	60 min or see Table B4 (whichever is greater)	60 min or see Table B4 (whichever is greater)	60 min or see Table B4 (whichever is greater)	From underside[5]
b. in upper storey of two storey dwellinghouse (but not over garage or basement)	R 30 and REI 15	30 min	15 min	15 min	From underside[5]
c. any other floor – including compartment floors.	REI see Table B4	See Table B4	See Table B4	See Table B4	From underside[5]
4. **Roofs**					
a. any part forming an escape route	REI 30	30 min	30 min	30 min	From underside[5]
b. any roof that performs the function of a floor.	REI see Table B4	See Table B4	See Table B4	See Table B4	From underside[5]

Table B3 Continued

Part of building	Minimum provisions when tested to the relevant European standard (minutes)[1]	Alternative minimum provisions when tested to the relevant part of **BS 476**[2] (minutes)			Type of exposure
		Loadbearing capacity[3]	Integrity	Insulation	
5. External walls					
a. any part a maximum of 1000mm from any point on the relevant boundary[6]	REI see Table B4	See Table B4	See Table B4	See Table B4	Each side separately
b. any part a minimum of 1000mm from the relevant boundary[6]	RE see Table B4 and REI 15	See Table B4	See Table B4	15 min	From inside the building
c. any part beside an external escape route (Section 2, Diagram 2.7 and Section 3, Diagram 3.11).	RE 30	30 min	30 min	No provision[7] [8]	From inside the building
6. Compartment walls Separating either:					
a. a flat from any other part of the building (see paragraph 7.1)	REI 60 or see Table B4 (whichever is less)	60 min or see Table B4 (whichever is less)	60 min or see Table B4 (whichever is less)	60 min or see Table B4 (whichever is less)	Each side separately
b. occupancies.	REI 60 or see Table B4 (whichever is less)	60 min or see Table B4 (whichever is less)	60 min or see Table B4 (whichever is less)	60 min or see Table B4 (whichever is less)	Each side separately
7. Compartment walls (other than in item 6 or item 10).	REI see Table B4	See Table B4	See Table B4	See Table B4	Each side separately
8. Protected shafts Excluding any firefighting shaft:					
a. any glazing	E 30	Not applicable	30 min	No provision[8]	Each side separately
b. any other part between the shaft and a protected lobby/corridor	REI 30	30 min	30 min	30 min	Each side separately
c. any part not described in (a) or (b) above.	REI see Table B4	See Table B4	See Table B4	See Table B4	Each side separately

B

Table B3 Continued

Part of building	Minimum provisions when tested to the relevant European standard (minutes)[1]	Alternative minimum provisions when tested to the relevant part of **BS 476**[2] (minutes)			Type of exposure
		Loadbearing capacity[3]	Integrity	Insulation	
9. **Enclosure** (that does not form part of a compartment wall or a protected shaft) to a:					
a. protected stairway	REI 30[8]	30 min	30 min	30 min[8]	Each side separately
b. lift shaft.	REI 30	30 min	30 min	30 min	Each side separately
10. **Wall or floor** separating an attached or integral garage from a dwellinghouse	REI 30[8]	30 min	30 min	30 min[8]	From garage side
11. **Fire resisting construction in dwellinghouses** not described elsewhere	REI 30[8]	30 min	30 min	30 min[8]	Each side separately
12. **Firefighting shafts**	REI 120	120 min	120 min	120 min	From side remote from shaft
a. construction that separates firefighting shaft from rest of building	REI 60	60 min	60 min	60 min	From shaft side
b. construction that separates firefighting stair, firefighting lift shaft and firefighting lobby.	REI 60	60 min	60 min	60 min	Each side separately
13. **Enclosure** (that is not a compartment wall or described in item 8) to a:					
a. protected lobby	REI 30[8]	30 min	30 min	30 min[8]	Each side separately
b. protected corridor.	REI 30[8]	30 min	30 min	30 min[8]	Each side separately
14. **Sub-division of a corridor**	REI 30[8]	30 min	30 min	30 min[8]	Each side separately

Table B3 Continued

Part of building	Minimum provisions when tested to the relevant European standard (minutes)[1]	Alternative minimum provisions when tested to the relevant part of **BS 476**[2] (minutes)			Type of exposure
		Loadbearing capacity[3]	Integrity	Insulation	
15. **Fire resisting construction**					
a. construction that encloses places of special fire hazard	REI 30	30 min	30 min	30 min	Each side separately
b. construction between store rooms and sales area in shops	REI 30	30 min	30 min	30 min	Each side separately
c. fire resisting sub-division	REI 30	30 min	30 min	30 min	Each side separately
d. construction that encloses bedrooms and ancillary accommodation in care homes.	REI 30	30 min	30 min	30 min	Each side separately
16. **Enclosure** in a flat to a protected entrance hall, or to a protected landing.	REI 30[8]	30 min	30 min	30 min[8]	Each side separately
17. **Cavity barrier**	E 30 and EI 15	Not applicable	30 min	15 min	Each side separately
18. **Ceiling** see paragraph 2.5 and Diagram 2.3; paragraph 8.5 and Diagram 8.3.	EI 30	Not applicable	30 min	30 min	From underside
19. **Duct** described in paragraph 5.24e.	E 30	Not applicable	30 min	No provision	From outside
20. **Casing** around a drainage system described in Diagram 9.1.	E 30	Not applicable	30 min	No provision	From outside
21. **Flue walls** described in Diagram 9.5.	EI half the period given in Table B4 for the compartment wall/floor	Not applicable	Half the period given in Table B4 for the compartment wall/floor	Half the period given in Table B4 for the compartment wall/floor	From outside

Table B3 Continued

Part of building	Minimum provisions when tested to the relevant European standard (minutes)[1]	Alternative minimum provisions when tested to the relevant part of **BS 476**[2] (minutes)			Type of exposure
		Loadbearing capacity[3]	Integrity	Insulation	
22. **Construction** described in note (a) to paragraph 12.9.	EI 30	Not applicable	30 min	30 min	From underside
23. **Fire doorsets**	See Table C1	See Table C1			See Appendix C

NOTES:

1. **BS EN 13501-2** Classification using data from fire resistance tests, excluding ventilation services. **BS EN 13501-3** Classification using data from fire resistance tests on products and elements used in building service installations: fire resisting ducts and fire dampers. **BS EN 13501-4** Classification using data from fire resistance tests on components of smoke control systems.

   In the European classification:

   'R' is the resistance to fire in terms of loadbearing capacity.

   'E' is the resistance to fire in terms of integrity.

   'I' is the resistance to fire in terms of insulation.

   The national classifications do not automatically equate with the alternative classifications in the European column, therefore products cannot typically assume a European class unless they have been tested accordingly.

2. **BS 476-20** for general principles, **BS 476-21** for loadbearing elements, **BS 476-22** for non-loadbearing elements, **BS 476-23** for fire-protecting suspended ceilings and **BS 476-24** for ventilation ducts.

3. Applies to loadbearing elements only (see paragraph B19).

4. Guidance on increasing the fire resistance of existing timber floors is given in BRE Digest 208.

5. Only if a suspended ceiling meets the appropriate provisions should it be relied on to add to the fire resistance of the floor.

6. Such walls may contain areas that do not need to be fire resisting (unprotected areas). See Section 11.

7. Unless needed as part of a wall in item 5a or 5b.

8. Except for any limitations on uninsulated glazed elements given in Table B5.

Table B4 Minimum periods of fire resistance

Purpose group of building	Minimum periods of fire resistance[1] (minutes) in a:					
	Basement storey* including floor over		Ground or upper storey			
	Depth (m) of the lowest basement		Height (m) of top floor above ground, in a building or separated part of a building			
	More than 10	Up to 10	Up to 5	Up to 18	Up to 30	More than 30
1. Residential:						
a. Block of flats						
– without sprinkler system	90 min	60 min	30 min[†]	60 min[+§]	90 min[+]	Not permitted[2]
– with sprinkler system[3]	90 min	60 min	30 min[†]	60 min[+§]	90 min[+]	120 min[+]
b. and c. Dwellinghouse	Not applicable[4]	30 min[*†]	30 min[†]	60 min[5]	Not applicable[4]	Not applicable[4]
2. Residential:						
a. Institutional	90 min	60 min	30 min[†]	60 min	90 min	120 min[‡]
b. Other residential	90 min	60 min	30 min[†]	60 min	90 min	120 min[‡]
3. Office:						
– without sprinkler system	90 min	60 min	30 min[†]	60 min	90 min	Not permitted[6]
– with sprinkler system[3]	60 min	60 min	30 min[†]	30 min[†]	60 min	120 min[‡]
4. Shop and commercial:						
– without sprinkler system	90 min	60 min	60 min	60 min	90 min	Not permitted
– with sprinkler system[3]	60 min	60 min	30 min[†]	60 min	60 min	120 min[‡]
5. Assembly and recreation:						
– without sprinkler system	90 min	60 min	60 min	60 min	90 min	Not permitted
– with sprinkler system[3]	60 min	60 min	30 min[†]	60 min	60 min	120 min[‡]
6. Industrial:						
– without sprinkler system	120 min	90 min	60 min	90 min	120 min	Not permitted
– with sprinkler system[3]	90 min	60 min	30 min[†]	60 min	90 min	120 min[‡]
7. Storage and other non-residential:						
a. any building or part not described elsewhere:						
– without sprinkler system	120 min	90 min	60 min	90 min	120 min	Not permitted
– with sprinkler system[3]	90 min	60 min	30 min[†]	60 min	90 min	120 min[‡]

Table B4 Continued

Purpose group of building	Minimum periods of fire resistance[1] (minutes) in a:					
	Basement storey* including floor over		Ground or upper storey			
	Depth (m) of the lowest basement		Height (m) of top floor above ground, in a building or separated part of a building			
	More than 10	Up to 10	Up to 5	Up to 18	Up to 30	More than 30
b. car park for light vehicles:						
i. open sided car park[7]	Not applicable	Not applicable	15 min[†#]	15 min[†#(8)]	15 min[†#(8)]	60 min
ii. any other car park	90 min	60 min	30 min[†]	60 min	90 min	120 min[‡]

NOTES:

For single storey buildings, the periods under the heading 'Up to 5' apply. If single storey buildings have basements, for the basement storeys the period appropriate to their depth applies.

* For the floor over a basement or, if there is more than one basement, the floor over the topmost basement, the higher of the period for the basement storey and the period for the ground or upper storey applies.

† For compartment walls that separate buildings, the period is increased to a minimum of 60 minutes.

+ For any floor that does not contribute to the support of the building within a flat of more than one storey, the period is reduced to 30 minutes.

§ For flat conversions, refer to paragraphs 6.5 to 6.7 regarding the acceptability of 30 minutes.

‡ For elements that do not form part of the structural frame, the period is reduced to 90 minutes.

For elements that protect the means of escape, the period is increased to 30 minutes.

1. Refer to note 1, Table B3 for the specific provisions of test.

2. Blocks of flats with a floor more than 30m above ground level should be fitted with a sprinkler system in accordance with Appendix E.

   NOTE: Sprinklers only need to be provided within the individual flats, they are not required in the common areas such as stairs, corridors or landings when these areas are fire sterile.

3. 'With sprinkler system' means that the building is fitted throughout with an automatic sprinkler system in accordance with Appendix E.

4. Very large (over 18m in height or with a 10m deep basement) or unusual dwellinghouses are outside the scope of the guidance provided with regard to dwellinghouses.

5. A minimum of 30 minutes in the case of three storey dwellinghouses, increased to 60 minutes minimum for compartment walls separating buildings.

6. Buildings within the 'office', 'shop and commercial', 'assembly and recreation', 'industrial' and 'storage and other non-residential' (except car parks for light vehicles) purpose groups (purpose groups 3 to 7(a)) require sprinklers where there is a top storey above 30m. The sprinkler system should be provided in accordance with Appendix E.

7. The car park should comply with the relevant provisions in the guidance on requirement B3, Section 11 of Approved Document B Volume 2.

8. For the purposes of meeting the Building Regulations, the following types of steel elements are deemed to have satisfied the minimum period of fire resistance of 15 minutes when tested to the European test method.

   i. Beams supporting concrete floors, maximum $Hp/A=230m^{-1}$ operating under full design load.

   ii. Free-standing columns, maximum $Hp/A=180m^{-1}$ operating under full design load.

   iii. Wind bracing and struts, maximum $Hp/A=210m^{-1}$ operating under full design load.

   Guidance is also available in **BS EN 1993-1-2**.

Application of the fire resistance standards in Table B4

B26 The following guidance should be used when applying the fire resistance standards in Table B4.

a. If one element of structure supports or carries or gives stability to another, the fire resistance of the supporting element should be no less than the minimum period of fire resistance for the other element (whether that other element is loadbearing or not). In some circumstances, it may be reasonable to vary this principle, for example:

   i. if the supporting structure is in the open air and is not likely to be affected by the fire in the building

   ii. if the supporting structure is in a different compartment, with a fire-separating element (that has the higher standard of fire resistance) between the supporting and the separated structure

   iii. if a plant room on the roof needs greater fire resistance than the elements of structure that support it.

b. If an element of structure forms part of more than one building or compartment, that element should be constructed to the standard of the higher of the relevant provisions.

c. If, due to the slope of the ground, one side of a basement is open at ground level (allowing smoke to vent and providing access for firefighting) for elements of structure in that storey it may be appropriate to adopt the standard of fire resistance that applies to above-ground structures.

d. Although most elements of structure in a single storey building may not need fire resistance, fire resistance is needed if one of the following applies to the element.

   i. It is part of, or supports, an external wall, and there is provision in the guidance on requirement B4 to limit the extent of openings and other unprotected areas in the wall.

   ii. It is part of, or supports, a compartment wall, including a wall that is common to two or more buildings.

   iii. It supports a gallery.

B27 For the purposes of this paragraph, the ground storey of a building that has one or more basement storeys and no upper storeys may be considered as a single storey building. The fire resistance of the basement storeys should be that specified for basements.

Table B5 Limitations on the use of uninsulated glazed elements on escape routes. These limitations *do not* apply to glazed elements that satisfy the relevant insulation criterion, see Table B3

Position of glazed element	Maximum total glazed area in parts of a building with access to:			
	A single stair		More than one stair	
	Walls	Door leaf	Walls	Door leaf
Flats				
1. Within the enclosures of a protected entrance hall or protected landing, or within fire resisting separation shown in Section 3, Diagram 3.4.	Fixed fanlights only	Unlimited above 1100mm from floor	Fixed fanlights only	Unlimited above 1100mm from floor
Dwellinghouses				
2. Within either: a. the enclosures of a protected stairway b. fire resisting separation shown in Diagram 2.2.	Unlimited above 1100mm from floor or pitch of the stair	Unlimited	Unlimited above 1100mm from floor or pitch of the stair	Unlimited
3. Within fire resisting separation either: a. shown in Diagram 2.4 b. described in paragraph 2.16b.	Unlimited above 100mm from floor	Unlimited above 100mm from floor	Unlimited above 100mm from floor	Unlimited above 100mm from floor
4. Existing window between an attached/integral garage and the dwellinghouse.	Unlimited	Not applicable	Unlimited	Not applicable
5. Adjacent to an external escape stair (see paragraph 2.17 and Diagram 2.7) or roof escape route (see paragraph 2.13).	Unlimited	Unlimited	Unlimited	Unlimited
General (except dwellinghouses)				
6. Between residential/sleeping accommodation and a common escape route (corridor, lobby or stair).	Nil	Nil	Nil	Nil
7. Between a protected stairway[1] and either: a. the accommodation b. a corridor that *is not* a protected corridor *other than in item 6 above*.	Nil	25% of door area	Unlimited above 1100mm[2]	50% of door area
8. Between either: a. a protected stairway[1] and a protected lobby or protected corridor b. accommodation and a protected lobby *other than in item 6 above*.	Unlimited above 1100mm from floor	Unlimited above 100mm from floor	Unlimited above 100mm from floor	Unlimited above 100mm from floor
9. Between the accommodation and a protected corridor that forms a dead end, *other than in item 6 above*.	Unlimited above 1100mm from floor	Unlimited above 100mm from floor	Unlimited above 1100mm from floor	Unlimited above 100mm from floor
10. Between accommodation and any other corridor, or sub-dividing corridors, *other than in item 6 above*.	Not applicable	Not applicable	Unlimited above 100mm from floor	Unlimited above 100mm from floor
11. Beside an external escape route.	Unlimited above 1100mm from floor	Unlimited above 1100mm from floor	Unlimited above 1100mm from floor	Unlimited above 1100mm from floor

Table B5 Continued

Position of glazed element	Maximum total glazed area in parts of a building with access to:			
	A single stair		More than one stair	
	Walls	Door leaf	Walls	Door leaf
12. Beside an external escape stair (see paragraph 3.68 and Diagram 3.11) or roof escape route (see paragraph 3.30).	Unlimited	Unlimited	Unlimited	Unlimited

NOTES:

Items 1 and 8 apply also to single storey buildings.

Fire resisting glass should be marked with the name of the manufacturer and the name of the product.

Further guidance can be found in *A Guide to Best Practice in the Specification and Use of Fire-resistant Glazed Systems* published by the Glass and Glazing Federation.

1. If the protected stairway is also a protected shaft or a firefighting stair (see Section 15), there may be further restrictions on the use of glazed elements.
2. Measured vertically from the landing floor level or the stair pitch line.
3. The 100mm limit is intended to reduce the risk of fire spreading from a floor covering.

C

Appendix C: Fire doorsets

C1 All fire doorsets should have the performance shown in Table C1, based on one of the following.

    a. Fire resistance in terms of integrity, for a period of minutes, when tested to **BS 476-22**, e.g. FD 30. A suffix (S) is added for doorsets where restricted smoke leakage at ambient temperatures is needed.

    b. As determined with reference to Commission Decision 2000/367/EC regarding the classification of the resistance to fire performance of construction products, construction works and parts thereof. All fire doorsets should be classified in accordance with **BS EN 13501-2**, tested to the relevant European method from the following.

        i.   **BS EN 1634-1**.

        ii.   **BS EN 1634-2**.

        iii.   **BS EN 1634-3**.

    c. As determined with reference to European Parliament and Council Directive 95/16/EC (which applies to lifts that permanently serve buildings and constructions and specified safety components) on the approximation of laws of Member States relating to lifts ('Lifts Directive') implementing the Lifts Regulations 1997 (SI 1997/831) and calling upon the harmonised standard **BS EN 81-58**.

C2 The performance requirement is in terms of integrity (E) for a period of minutes. An additional classification of S_a is used for all doors where restricted smoke leakage at ambient temperatures is needed.

C3 The requirement is for test exposure from each side of the doorset separately. The exception is lift doors, which are tested from the landing side only.

C4 Any test evidence used to verify the fire resistance rating of a doorset or shutter should be checked to ensure both of the following.

    a. It adequately demonstrates compliance.

    b. It is applicable to the **complete installed assembly**. Small differences in detail may significantly affect the rating.

Until relevant harmonised product standards are published, for the purposes of meeting the Building Regulations, products tested in accordance with **BS EN 1634-1** (with or without pre-fire test mechanical conditioning) that achieve the minimum performance in Table C1 will be deemed to satisfy the provisions.

C5 All fire doorsets, including to flat entrances and between a dwellinghouse and an integral garage, should be fitted with a self-closing device, except for all of the following.

    a. Fire doorsets to cupboards.

    b. Fire doorsets to service ducts normally locked shut.

    c. Fire doorsets within flats and dwellinghouses.

C6 If a self-closing device would be considered to interfere with the normal approved use of the building, self-closing fire doors may be held open by one of the following.

    a. A fusible link, but not if the doorset is in an opening provided as a means of escape unless it complies with paragraph C7.

    b. An automatic release mechanism activated by an automatic fire detection and alarm system.

    c. A door closer delay device.

C7 Two fire doorsets may be fitted in the same opening if each door is capable of closing the opening, so the total fire resistance is the sum of their individual resistances. If the opening is provided as a means of escape, both fire doorsets should be self-closing.

If one fire doorset is capable of being easily opened by hand and has a minimum of 30 minutes' fire resistance, the other fire doorset should comply with both of the following.

    a. Be fitted with an automatic self-closing device.

    b. Be held open by a fusible link.

C8 Fire doorsets often do not provide any significant insulation. Unless providing both integrity and insulation in accordance with Appendix B, Table B3, a maximum of 25% of the length of a compartment wall should consist of door openings.

Where it is practicable to maintain a clear space on both sides of the doorway, the above percentage may be greater.

C9 Rolling shutters should be capable of manual opening and closing for firefighting purposes (see Section 15). Rolling shutters across a means of escape should only be released by a heat sensor, such as a fusible link or electric heat detector, in the immediate vicinity of the door.

Unless a shutter is also intended to partially descend as part of a boundary to a smoke reservoir, shutters across a means of escape should not be closed by smoke detectors or a fire alarm system.

C10 Unless shown to be satisfactory when tested as part of a fire doorset assembly, the essential components of any hinge on which a fire door is hung should be made entirely from materials that have a minimum melting point of 800°C.

C11 Except for doorsets listed in paragraph C12, all fire doorsets should be marked with one of the following fire safety signs, complying with **BS 5499-5**, as appropriate.

    a. To be kept closed when not in use – mark 'Fire door keep shut'.

    b. To be kept locked when not in use – mark 'Fire door keep locked shut'.

    c. Held open by an automatic release mechanism or free swing device – mark 'Automatic fire door keep clear'.

All fire doorsets should be marked on both sides, except fire doorsets to cupboards and service ducts, which should be marked on the outside.

C12 The following fire doorsets are not required to comply with paragraph C11.

    a. Doors to and within flats and dwellinghouses.

    b. Bedroom doors in 'residential (other)' (purpose group 2(b)) premises.

    c. Lift entrance/landing doors.

C13 The performance of some doorsets set out in Table C1 is linked to the minimum periods of fire resistance for elements of structure given in Tables B3 and B4. Limitations on the use of uninsulated glazing in fire doorsets are given in Table B5.

C14 Recommendations for the specification, design, construction, installation and maintenance of fire doorsets constructed with non-metallic door leaves are given in **BS 8214**.

Guidance on timber fire resisting doorsets, in relation to the new European test standard, may be found in *Timber Fire Resisting Doorsets: Maintaining Performance Under the New European Test Standard* published by the Timber Research and Development Association (TRADA).

Guidance for metal doors is given in *Code of Practice for Fire Resisting Metal Doorsets* published by the Door and Shutter Manufacturers' Association (DSMA).

C15 Hardware used on fire doors can significantly affect their performance in a fire. Notwithstanding the guidance in this approved document, guidance is available in *Hardware for Fire and Escape Doors* published by the Door and Hardware Federation (DHF) and Guild of Architectural Ironmongers (GAI).

Table C1 Provisions for fire doorsets

Position of door	Minimum fire resistance of door in terms of integrity (minutes) when tested to the relevant European standard[1]	Minimum fire resistance of door in terms of integrity (minutes) when tested to **BS 476-22**
1. **In a compartment wall separating buildings**	Same as for the wall in which the door is fitted, but a minimum of 60 minutes	Same as for the wall in which the door is fitted, but a minimum of 60 minutes
2. **In a compartment wall:**		
a. if it separates a flat from a space in common use	E 30 S$_a$[2]	FD 30 S[2]
b. enclosing a protected shaft forming a stairway wholly or partly above the adjoining ground in a building used for flats, other residential, assembly and recreation, or office purposes	E 30 S$_a$[2]	FD 30 S[2]
c. enclosing a protected shaft forming a stairway not described in (b) above	Half the period of fire resistance of the wall in which it is fitted, but 30 minutes minimum and with suffix S$_a$[2]	Half the period of fire resistance of the wall in which it is fitted, but 30 minutes minimum and with suffix S[2]
d. enclosing a protected shaft forming a lift or service shaft	Half the period of fire resistance of the wall in which it is fitted, but 30 minutes minimum	Half the period of fire resistance of the wall in which it is fitted, but 30 minutes minimum
e. not described in (a), (b), (c) or (d) above.	Same as for the wall in which it is fitted, but add S$_a$[2] if the door is used for progressive horizontal evacuation under the guidance to requirement B1	Same as for the wall in which it is fitted, but add S[2] if the door is used for progressive horizontal evacuation under the guidance to requirement B1
3. **In a compartment floor**	Same as for the floor in which it is fitted	Same as for the floor in which it is fitted

Table C1 Continued

Position of door	Minimum fire resistance of door in terms of integrity (minutes) when tested to the relevant European standard[1]	Minimum fire resistance of door in terms of integrity (minutes) when tested to **BS 476-22**
4. Forming part of the enclosures of:		
a. a protected stairway (except as described in item 9 or 11(b) below)	E 30 S_a[2]	FD 30 S[2]
b. a lift shaft (see paragraph 3.99b) that does not form a protected shaft in 2(b), (c) or (d) above.	E 30	FD 30
5. Forming part of the enclosure of:		
a. a protected lobby approach (or protected corridor) to a stairway	E 30 S_a[2]	FD 30 S[2]
b. any other protected corridor	E 20 S_a[2]	FD 20 S[2]
c. a protected lobby approach to a lift shaft (paragraphs 3.102 to 3.104).	E 30 S_a[2]	FD 30 S[2]
6. Giving access to an external escape route	E 30	FD 30
7. Sub-dividing:		
a. corridors connecting alternative exits	E 20 S_a[2]	FD 20 S[2]
b. dead-end portions of corridors from the remainder of the corridor.	E 20 S_a[2]	FD 20 S[2]
8. Any door within a cavity barrier	E 30	FD 30
9. Any door that forms part of the enclosure to a protected entrance hall or protected landing in a flat	E 20	FD 20
10. Any door that forms part of the enclosure:		
a. to a place of special fire hazard	E 30	FD 30
b. to ancillary accommodation in care homes (see paragraph 2.44 in Approved Document B Volume 2).	E 30	FD 30
11. In a dwellinghouse:		
a. between a dwellinghouse and a garage	E 30 S_a[2]	FD 30 S[2]
b. forming part of the enclosures to a protected stairway in a single family dwellinghouse	E 20	FD 20
c. within any fire resisting construction in a dwellinghouse not described elsewhere in this table.	E 20	FD 20

NOTES:

1. Classified in accordance with **BS EN 13501-2**. National classifications do not necessarily equate with European classifications, therefore products cannot typically assume a European class unless they have been tested accordingly.

2. Unless pressurisation techniques that comply with **BS EN 12101-6** are used, these doors should also comply with one of the following conditions.

    a. Have a leakage rate not exceeding 3m³/m/hour (from head and jambs only) when tested at 25Pa under **BS 476-31.1**.

    b. Meet the additional S_a classification when tested to **BS EN 1634-3**.

Appendix D: Methods of measurement

Occupant number

D1 The number of occupants of a room, storey, building or part of a building is either of the following.

    a. The maximum number of people it is designed to hold.

    b. In buildings other than dwellings, the number of people calculated by dividing the area of a room or storey(s) (m^2) by a floor space factor (m^2 per person) such as given in Table D1 for guidance.

D2 Counters and display units should be included when measuring area. All of the following should be *excluded*.

    a. Stair enclosures.

    b. Lifts.

    c. Sanitary accommodation.

    d. Any other fixed part of the building structure.

Table D1 Floor space factors[1]

Type of accommodation[2][3]	Floor space factor (m²/person)
1. Standing spectator areas, bar areas (within 2m of serving point), similar refreshment areas	0.3
2. Amusement arcade, assembly hall (including a general purpose place of assembly), bingo hall, club, crush hall, dance floor or hall, venue for pop concerts and similar events and bar areas without fixed seating	0.5
3. Concourse or queuing area[4]	0.7
4. Committee room, common room, conference room, dining room, licensed betting office (public area), lounge or bar (other than in (1) above), meeting room, reading room, restaurant, staff room or waiting room[5]	1.0
5. Exhibition hall or studio (film, radio, television, recording)	1.5
6. Skating rink	2.0
7. Shop sales area[6]	2.0
8. Art gallery, dormitory, factory production area, museum or workshop	5.0
9. Office	6.0
10. Shop sales area[7]	7.0
11. Kitchen or library	7.0
12. Bedroom or study-bedroom	8.0
13. Bed-sitting room, billiards or snooker room or hall	10.0
14. Storage and warehousing	30.0
15. Car park	Two persons per parking space

NOTES:

1. As an alternative to using the values in the table, the floor space factor may be determined by reference to actual data taken from similar premises. Where appropriate, the data should reflect the average occupant density at a peak trading time of year.

2. Where accommodation is not directly covered by the descriptions given, a reasonable value based on a similar use may be selected.

3. Where any part of the building is to be used for more than one type of accommodation, the most onerous factor(s) should be applied. Where the building contains different types of accommodation, the occupancy of each different area should be calculated using the relevant space factor.

4. For detailed guidance on appropriate floor space factors for concourses in sports grounds refer to *Concourses* published by the Football Licensing Authority.

5. Alternatively the occupant number may be taken as the number of fixed seats provided, if the occupants will normally be seated.

6. Shops excluding those under item 10, but including: supermarkets and department stores (main sales areas), shops for personal services, such as hairdressing, and shops for the delivery or collection of goods for cleaning, repair or other treatment or for members of the public themselves carrying out such cleaning, repair or other treatment.

7. Shops (excluding those in covered shopping complexes but including department stores) trading predominantly in furniture, floor coverings, cycles, prams, large domestic appliances or other bulky goods, or trading on a wholesale self-selection basis (cash and carry).

Travel distance

D3 Travel distance is measured as the shortest route. Both of the following should be observed.

    a. If there is fixed seating or other fixed obstructions, the shortest route is along the centre line of the seatways and gangways.

    b. If the route includes a stair, the shortest route is along the pitch line on the centre line of travel.

Width

D4 Width is measured according to the following.

    a. For a **door (or doorway)**, the clear width when the door is open (Diagram D1).

    b. For an escape route, either of the following.

        i.   When the route is defined by walls: the width at 1500mm above finished floor level.

        ii.   Elsewhere: the minimum width of passage available between any fixed obstructions.

    c. For a **stair**, the clear width between the walls or balustrades. On escape routes and stairs, handrails and strings intruding into the width by a maximum of 100mm on each side may be ignored. Rails used for guiding a stair-lift may be ignored, but it should be possible to park the lift's chair or carriage in a position that does not obstruct the stair or landing.

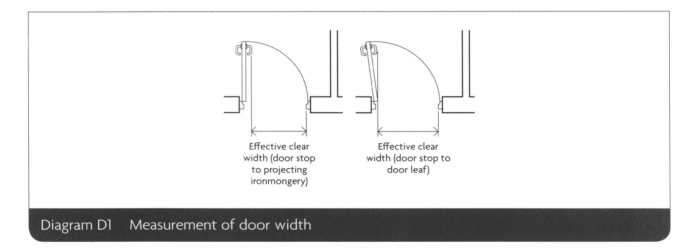

Effective clear width (door stop to projecting ironmongery)

Effective clear width (door stop to door leaf)

Diagram D1   Measurement of door width

Building dimensions

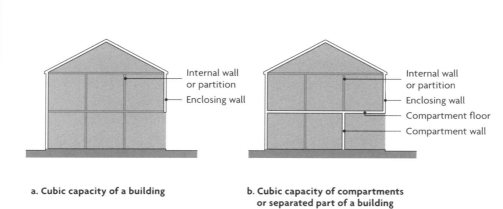

In every case measure the volume contained by all of the following.

a. Under surface of roof.

b. Upper surface of lowest floor.

c. Inner surface of enclosing walls. When there is not an outer enclosing wall, measure to the outermost edge of the floor slab.

The measured volume should include internal walls and partitions.

a. Cubic capacity of a building

b. Cubic capacity of compartments or separated part of a building

Diagram D2   Cubic capacity

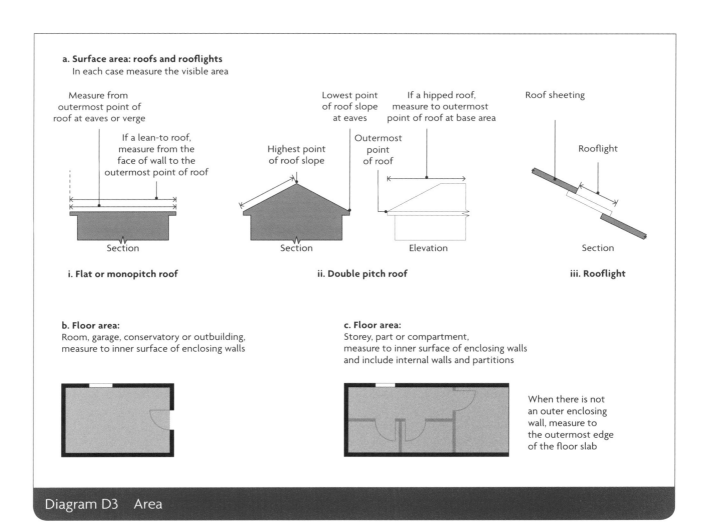

a. Surface area: roofs and rooflights
In each case measure the visible area

Measure from outermost point of roof at eaves or verge

If a lean-to roof, measure from the face of wall to the outermost point of roof

Lowest point of roof slope at eaves

If a hipped roof, measure to outermost point of roof at base area

Highest point of roof slope

Outermost point of roof

Roof sheeting

Rooflight

Section

Section

Elevation

Section

i. Flat or monopitch roof

ii. Double pitch roof

iii. Rooflight

b. Floor area:
Room, garage, conservatory or outbuilding, measure to inner surface of enclosing walls

c. Floor area:
Storey, part or compartment, measure to inner surface of enclosing walls and include internal walls and partitions

When there is not an outer enclosing wall, measure to the outermost edge of the floor slab

Diagram D3   Area

D

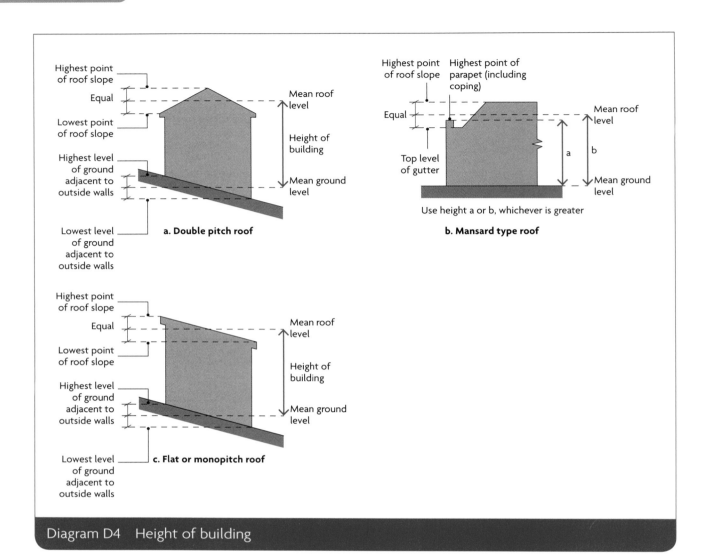

Diagram D4   Height of building

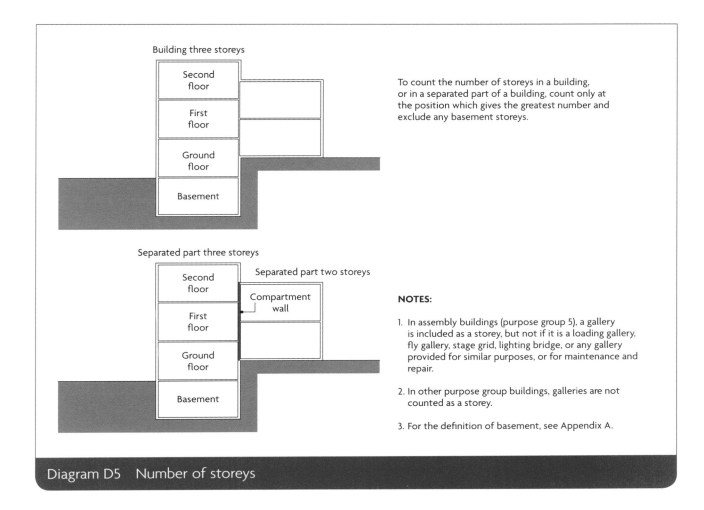

To count the number of storeys in a building, or in a separated part of a building, count only at the position which gives the greatest number and exclude any basement storeys.

NOTES:

1. In assembly buildings (purpose group 5), a gallery is included as a storey, but not if it is a loading gallery, fly gallery, stage grid, lighting bridge, or any gallery provided for similar purposes, or for maintenance and repair.

2. In other purpose group buildings, galleries are not counted as a storey.

3. For the definition of basement, see Appendix A.

Diagram D5   Number of storeys

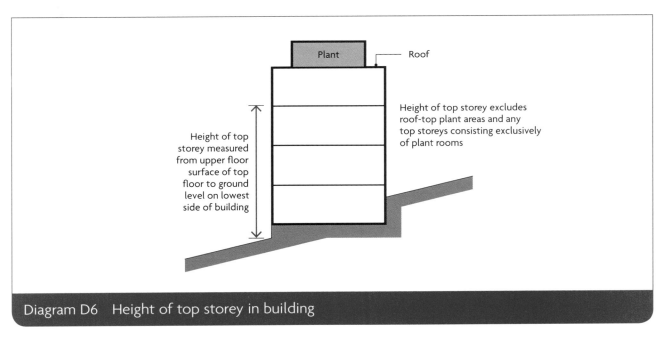

Height of top storey excludes roof-top plant areas and any top storeys consisting exclusively of plant rooms

Diagram D6   Height of top storey in building

Free area of smoke ventilators

D5 The free area of a smoke ventilator should be measured by either of the following.

    a. The declared aerodynamic free area in accordance with **BS EN 12101-2**.

    b. The total unobstructed cross-sectional area (geometric free area), measured in the plane where the area is at a minimum and at right angles to the direction of air flow (Diagram D7).

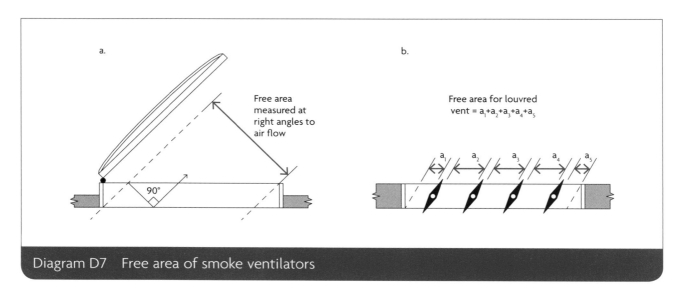

Diagram D7   Free area of smoke ventilators

Appendix E: Sprinklers

Sprinkler systems

E1   Sprinkler systems installed in buildings can reduce the risk to life and significantly reduce the degree of damage caused by fire within a building.

E2   Further recommendations for the provision of sprinklers are provided in the following sections:

Volume 1 – Dwellings

Functional requirement	Paragraph	Title
B1	2.6	Dwellinghouses with two or more storeys more than 4.5m above ground level
B1	2.23	Loft conversions
B1	3.21	Internal planning of multi-storey flats
B3	Table B4	Minimum periods of fire resistance
B3	7.4	Sprinklers
B4	11.15	Unprotected areas and fire resistance – portal frames
B4	11.21	Methods for calculating acceptable unprotected area – sprinkler systems
B5	15.7	Provision of firefighting shafts

Volume 2 – Buildings other than dwellings

Functional requirement	Paragraph	Title
B1	2.46	Residential care homes – sprinkler systems
B1	3.21	Width of escape stairs – phased evacuation
B1	5.46	Shop store rooms
B3	7.7	Raised storage areas
B3	Table 8.1	Maximum dimensions of building or compartment
B3	Table B4	Minimum periods of fire resistance
B3	8.14	Sprinklers
B4	13.16	Unprotected areas and fire resistance – portal frames
B4	13.22	Methods for calculating acceptable unprotected area – sprinkler systems
B5	17.8	Location of firefighting shafts
B5	18.11	Provision of smoke outlets – mechanical smoke extract

Design of sprinkler systems

E3 Where required, sprinkler systems should be provided throughout the building or separated part, unless acting as a compensatory feature to address a specific risk. They should be designed and installed in accordance with the following.

a. For residential buildings, the requirements of **BS 9251**.

b. For non-residential buildings, or residential buildings outside the scope of **BS 9251**, the requirements of **BS EN 12845**, including the relevant hazard classification together with additional measures to improve system reliability and availability as described in Annex F of the standard.

NOTE: Any sprinkler system installed to satisfy the requirements of Part B of the Building Regulations should be provided with additional measures to improve system reliability and availability and is therefore to be regarded as a life safety system. However, there may be some circumstances in which additional measures to improve system reliability and availability specified in Annex F of **BS EN 12845** are inappropriate or unnecessary.

E4 If the provisions in a building vary from those in this document, sprinkler protection can also sometimes be used as a compensatory feature.

BS 9251 makes additional recommendations when sprinklers are proposed as compensatory features.

Water supplies and pumps

E5 For non-residential sprinkler systems designed and installed to **BS EN 12845**, water supplies should consist of either of the following.

a. Two single water supplies complying with clause 9.6.1, independent of each other.

b. Two stored water supplies meeting all of the following conditions.

   i. Gravity or suction tanks should satisfy all the requirements of clause 9.6.2(b), other than capacity.

   ii. Any pump arrangements should comply with clause 10.2.

   iii. In addition to meeting the requirements for inflow, either of the following should apply.

   - The capacity of each tank should be at least half the specified minimum water volume of a single full capacity tank, appropriate to the hazard.

   - One tank should be at least equivalent to half the specified water volume of a single full capacity tank, and the other shall not be less than the minimum volume of a reduced capacity tank (see clause 9.3.4) appropriate to the hazard.

   The total capacity of the water supply in (iii), including any inflow for a reduced capacity tank, should be at least that of a single full holding capacity tank that complies with Table 9, Table 10 or clause 9.3.2.3, as appropriate to the hazard and pipework design.

E6 For the systems described in paragraph E5, both of the following apply if pumps are used to draw water from two tanks.

a. Each pump should be able to draw water from either tank.

b. Any one pump, or either tank, should be able to be isolated.

The sprinkler water supplies should not be used as connections for other services or other fixed firefighting systems.

Appendix F: Standards referred to

European Standards

NOTE: All the British and European Standards can be purchased at the following address: https://shop.bsigroup.com/. Alternatively access to the British and European Standards may be gained at public reference libraries.

BS EN 54 Fire detection and fire alarm systems

>**BS EN 54-7** Smoke detectors. Point smoke detectors that operate using scattered light, transmitted light or ionization [2018]

>**BS EN 54-11** Manual call points [2001]

BS EN 81 Safety rules for the construction and installation of lifts

>**BS EN 81-20** Lifts for the transport of persons and goods. Passenger and goods passenger lifts [2014]

>**BS EN 81-58** Examination and tests. Landing doors fire resistance test [2018]

>**BS EN 81-72** Particular applications for passenger and goods passenger lifts. Firefighters lifts [2015]

BS EN ISO 306 Plastics. Thermoplastic materials. Determination of Vicat softening temperature (VST) [2013]

BS EN 520 Gypsum plasterboards. Definitions, requirements and test methods [2004 + A1 2009]

BS EN 1125 Building hardware. Panic exit devices operated by a horizontal bar, for use on escape routes. Requirements and test methods [2008]

BS EN 1155 Building hardware. Electrically powered hold-open devices for swing doors. Requirements and test methods [1997]

BS EN 1366 Fire resistance tests for service installations

>**BS EN 1366-2** Fire dampers [2015]

>**BS EN 1366-8** Smoke extraction ducts [2004]

BS EN 1634 Fire resistance and smoke control tests for door and shutter assemblies, openable windows and elements of building hardware

>**BS EN 1634-1** Fire resistance test for door and shutter assemblies and openable windows [2014 + A1 2018]

>**BS EN 1634-2** Fire resistance characterisation test for elements of building hardware [2008]

>**BS EN 1634-3** Smoke control test for door and shutter assemblies [2004]

BS EN 1993-1-2 Eurocode 3. Design of steel structures. General rules. Structural fire design [2005]

BS ISO 3864-1 Graphical symbols. Safety colours and safety signs. Design principles for safety signs and safety markings [2011]

BS EN 12101 Smoke and heat control systems

>**BS EN 12101-2** Natural smoke and heat exhaust ventilators [2017]

>**BS EN 12101-3** Specification for powered smoke and heat control ventilators (Fans) [2015]

>**BS EN 12101-6** Specification for pressure differential systems. Kits [2005]

BS EN 12845 Fixed firefighting systems. Automatic sprinkler systems. Design, installation and maintenance [2015]

BS EN 13238 Reaction to fire tests for building products. Conditioning procedures and general rules for selection of substrates [2010]

BS EN 13501 Fire classification of construction products and building elements

>**BS EN 13501-1** Classification using data from reaction to fire tests [2018]

>**BS EN 13501-2** Classification using data from fire resistance tests, excluding ventilation services [2016]

BS EN 13501-3 Classification using data from fire resistance tests on products and elements used in building service installations: fire resisting ducts and fire dampers [2005 + A1 2009]

BS EN 13501-4 Classification using data from fire resistance tests on components of smoke control systems [2016]

BS EN 13501-5 Classification using data from external fire exposure to roof tests [2016]

BS EN 14306 Thermal insulation products for building equipment and industrial installations. Factory made calcium silicate (CS) products. Specification [2015]

BS EN 14604 Smoke alarm devices [2005]

BS EN 15102 Decorative wall coverings. Roll and panel form [2007 + A1 2011]

BS EN 15650 Ventilation for buildings. Fire dampers [2010]

BS EN 15725 Extended application reports on the fire performance of construction products and building elements [2010]

BS EN 50200 Method of test for resistance to fire of unprotected small cables for use in emergency circuits [2015]

British Standards

BS 476 Fire tests on building materials and structures

BS 476-3 Classification and method of test for external fire exposure to roofs [2004]

BS 476-6 Method of test for fire propagation for products [1989 + A1 2009]

BS 476-7 Method of test to determine the classification of the surface spread of flame of products [1997]

BS 476-8 Test methods and criteria for the fire resistance of elements of building construction [1972]

BS 476-11 Method for assessing the heat emission from building materials [1982]

BS 476-20 Method for determination of the fire resistance of elements of construction (general principles) [1987]

BS 476-21 Methods for determination of the fire resistance of loadbearing elements of construction [1987]

BS 476-22 Methods for determination of the fire resistance of non-loadbearing elements of construction [1987]

BS 476-23 Methods for determination of the contribution of components to the fire resistance of a structure [1987]

BS 476-24 Method for determination of the fire resistance of ventilation ducts [1987]

BS 476-31.1 Methods for measuring smoke penetration through doorsets and shutter assemblies. Method of measurement under ambient temperature conditions [1983]

BS 2782-0 Methods of testing. Plastics. Introduction [2011]

BS 3251 Specification. Indicator plates for fire hydrants and emergency water supplies [1976]

BS 4422 Fire. Vocabulary [2005]

BS 4514 Unplasticized PVC soil and ventilating pipes of 82.4mm minimum mean outside diameter, and fittings and accessories of 82.4mm and of other sizes. Specification [2001]

BS 5255 Specification for thermoplastics waste pipe and fittings [1989]

BS 5266-1 Emergency lighting. Code of practice for the emergency lighting of premises [2016]

BS 5395-2 Stairs, ladders and walkways. Code of practice for the design of helical and spiral stairs [1984]

BS 5438 Methods of test for flammability of textile fabrics when subjected to a small igniting flame applied to the face or bottom edge of vertically oriented specimens [1989]

BS 5446-2 Fire detection and fire alarm devices for dwellings. Specification for heat alarms [2003]

BS 5499 Graphical symbols and signs

> **BS 5499-4** Safety signs. Code of practice for escape route signing [2013]

> **BS 5499-5** Safety signs, including fire safety signs. Signs with specific safety meanings [2002]

BS 5839 Fire detection and fire alarm systems for buildings

> **BS 5839-1** Code of practice for system design, installation, commissioning and maintenance of systems in non-domestic premises [2017]

> **BS 5839-2** Specification for manual call points [1983]

> **BS 5839-3** Specification for automatic release mechanisms for certain fire protection equipment [1988]

> **BS 5839-6** Code of practice for the design, installation, commissioning and maintenance of fire detection and fire alarm systems in domestic premises [2019]

> **BS 5839-8** Code of practice for the design, installation, commissioning and maintenance of voice alarm systems [2013]

> **BS 5839-9** Code of practice for the design, installation, commissioning and maintenance of emergency voice communication systems [2011]

BS 5867-2 Fabrics for curtains and drapes. Flammability requirements. Specification [2008]

BS 5906 Waste management in buildings. Code of practice [2005]

BS 7157 Method of test for ignitability of fabrics used in the construction of large tented structures [1989]

BS 7273 Code of practice for the operation of fire protection measures

> **BS 7273-4** Actuation of release mechanisms for doors [2015]

BS 7346-7 Components for smoke and heat control systems. Code of practice on functional recommendations and calculation methods for smoke and heat control systems for covered car parks [2013]

BS 7974 Application of fire safety engineering principles to the design of buildings. Code of practice [2019]

BS 8214 Timber-based fire door assemblies. Code of practice [2016]

BS 8313 Code of practice for accommodation of building services in ducts [1997]

BS 8414 Fire performance of external cladding systems

> **BS 8414-1** Test method for non-loadbearing external cladding systems applied to the masonry face of a building [2015 + A1 2017]

> **BS 8414-2** Test method for non-loadbearing external cladding systems fixed to and supported by a structural steel frame [2015 + A1 2017]

BS 8519 Selection and installation of fire-resistant power and control cable systems for life safety and fire-fighting applications. Code of practice [2010]

BS 9251 Fire sprinkler systems for domestic and residential occupancies. Code of practice [2014]

BS 9252 Components for residential sprinkler systems. Specification and test methods for residential sprinklers [2011]

BS 9990 Non automatic fire-fighting systems in buildings. Code of practice [2015]

BS 9991 Fire safety in the design, management and use of residential buildings. Code of practice [2015]

BS 9999 Fire safety in the design, management and use of buildings. Code of practice [2017]

G

Appendix G: Documents referred to

Legislation

(available via www.legislation.gov.uk)

Education Act 1996

Gas Safety (Installation and Use) Regulations 1998 (SI 1998/2451)

Lifts Regulations 1997 (SI 1997/831)

Pipelines Safety Regulations 1996 (SI 1996/825)

Prison Act 1952

Safety of Sports Grounds Act 1975

Regulatory Reform (Fire Safety) Order 2005 (SI 2005/1541)

Commission Decision 2000/367/EC of 3 May 2000 implementing Council Directive 89/106/EEC

Commission Decision 2000/553/EC of 6 September 2000 implementing Council Directive 89/106/EEC

European Parliament and Council Directive 95/16/EC

Other documents

Publications

Association for Specialist Fire Protection (ASFP) (www.asfp.org.uk)

ASFP Red Book – *Fire-Stopping: Linear Joint Seals, Penetration Seals and Cavity Barriers,* Fourth Edition

ASFP Grey Book – *Volume 1: Fire Dampers (European Standards)*, Second Edition

ASFP Blue Book British Standard version – *Fire Resisting Ductwork, Tested to BS 476 Part 24,* Third Edition

ASFP Blue Book European version – *Fire Resisting Ductwork, Classified to BS EN 13501 Parts 3 and 4,* First Edition

Ensuring Best Practice for Passive Fire Protection in Buildings, Second Edition [2014]

Building Research Establishment Limited (BRE) (www.bre.co.uk)

BRE report (BR 135) *Fire Performance of External Thermal Insulation for Walls of Multi-storey Buildings*, Third Edition [2013]

BRE report (BR 187) *External Fire Spread: Building Separation and Boundary Distances*, Second Edition [2014]

BRE Digest 208 *Increasing the Fire Resistance of Existing Timber Floors* [1988]

BRE report (BR 274) *Fire Safety of PTFE-based Materials Used in Buildings* [1994]

Department for Communities and Local Government
(www.gov.uk/government/publications/fire-performance-of-green-roofs-and-walls)

Fire Performance of Green Roofs and Walls [2013]

Department for Education
(www.dfes.gov.uk)

Building Bulletin (BB) 100: *Design for Fire Safety in Schools* [2007]

Department of Health
(www.dh.gov.uk)

Health Technical Memorandum (HTM) 05-02: *Firecode. Guidance in Support of Functional Provisions (Fire Safety in the Design of Healthcare Premises)* [2015]

HTM 88: *Guide to Fire Precautions in NHS Housing in the Community for Mentally Handicapped (or Mentally Ill) People*

Door and Hardware Federation (DHF) and Guild of Architectural Ironmongers (GAI)
(www.firecode.org.uk)

Hardware for Fire and Escape Doors [2012]

Door and Shutter Manufacturers' Association (DSMA)
(www.dhfonline.org.uk)

Code of Practice for Fire Resisting Metal Doorsets [2010]

Fire Protection Association (FPA)
(www.thefpa.co.uk)

RISCAuthority Design Guide for the Fire Protection of Buildings [2005]

Football Licensing Authority
(www.flaweb.org.uk/home.php)

Concourses [2006]

Glass and Glazing Federation (GGF)
(www.ggf.org.uk)

A Guide to Best Practice in the Specification and Use of Fire-resistant Glazed Systems [2011]

Health and Safety Executive (HSE)
(www.hse.gov.uk)

Safety Signs and Signals: The Health and Safety Regulations 1996. Guidance on Regulations, L64 [2015]

HM Prison and Probation Service (HMPPS)
(www.hmppsintranet.org.uk/uploads/HMPPSFireSafetyDesignGuide.pdf)

Custodial Premises Fire Safety Design Guide

Passive Fire Protection Federation (PFPF)
(http://pfpf.org/pdf/publications/guide_to_uailoft.pdf)

Guide to Undertaking Assessments in Lieu of Fire Tests [2000]

Sports Grounds Safety Authority
(https://sgsa.org.uk/)

Guide to Safety at Sports Grounds [2007]

Steel Construction Institute (SCI)
(https://steel-sci.com)

SCI Publication P288 *Fire Safe Design: A New Approach to Multi-storey Steel-framed Buildings, Second Edition* [2006]

SCI Publication P313 *Single Storey Steel Framed Buildings in Fire Boundary Conditions* [2002]

Timber Research and Development Associations (TRADA)
(www.trada.co.uk)

Timber Fire Resisting Doorsets: Maintaining Performance under the New European Test Standard [2002]

Index

A

Access floors
See Platform floors
Access for fire service
See Fire service facilities
Accessibility 0.8, 3.107, 17.4
Access panels
Drainage and water supply pipes Diagram 9.1
Access rooms
Definition Appendix A
Means of escape from inner rooms 2.12, 3.8
Accreditation
Installers and suppliers page iii, Appendix B5
Air changes
See Ventilation
Air circulation systems for heating, etc.
Dwellinghouses with a floor more than 4.5m above ground level 2.8–2.9
Multi-storey flats 3.23
Air conditioning 9.6–9.17
See also Ventilation
Air exhaust terminals 9.6
Alarm systems
See Fire detection and alarm systems
Alterations
Material alteration 1.8–1.9
Alternative approaches 0.9
Alternative escape routes
Cavity barriers 8.6
Definition Appendix A
Dwellinghouses with floor more than 4.5m above ground level 2.5–2.6
Flats 3.26–3.27, 3.59
Alternative exits
Common balconies 3.12
Definition Appendix A
Divided corridors Table C1
Dwellinghouses with floor more than 4.5m above ground level Diagram 2.4
Flats
  Multi-storey flats 3.21, 3.22, Diagrams 3.5 to 3.6
  Single storey 3.19, Diagram 3.4
Galleries 2.15, 3.13

Alternative supply of water 14.12–14.13
Ancillary accommodation 3.28, 3.37, 3.73–3.75
Appliance ventilation ducts 9.23
Definition Appendix A
Approved Documents pages i–ii
Architraves
Definition of wall and ceiling 4.3, 4.6
Area measurement Diagram D3
Assembly and recreation purpose group
  Table 0.1
Atria 7.20
Definition Appendix A
Automatic doors 3.98
See also Automatic release mechanisms; Self-closing devices
Automatic fire detection and alarm systems
See Fire detection and alarm systems
Automatic release mechanisms
Definition Appendix A
Fire and smoke dampers 9.14, 9.22
Fire safety signs Appendix C11
Self-closing fire doors Appendix C6
See also Self-closing devices

B

Back gardens
Means of escape 2.10, Diagram 2.5
Balconies
Escape routes 3.29
Means of escape 2.14, 3.11–3.12
See also Common balconies
Basement storeys
Definition Appendix A
Escape stairs 3.82
Firefighting shafts 15.3
Fire resistance Appendix B27, Table B4
Lifts 3.102, 3.104
Means of escape 2.16, 3.9, 3.71–3.72
Smoke outlets 16.2–16.3
Bathrooms
Fire doorsets Diagram 3.2
Inner rooms 2.11, 3.7
Beams
Fire resistance Table B3
Openings for 5.9, 7.6

Boiler rooms 3.73
Boundaries
Definition Appendix A
See also Notional boundaries; Relevant boundaries; Separation distances
British Standards Appendix F
BS 476-3 Table B2
BS 476-6 Table B1
BS 476-7 Appendix B13, Table B1
BS 476-20 to 21 Table B3
BS 476-22 Table B3, Appendix C1, Table C1
BS 476-23 to 24 Table B3
BS 476-31.1 Table C1
BS 2782 Appendix B13
BS 2782-0 Appendix B13
BS 3251 14.10
BS 4422 Appendix A
BS 4514 9.5, Table 9.1
BS 5255 9.5, Table 9.1
BS 5266-1 3.24, 3.44, 3.48
BS 5395-2 3.86
BS 5438 Appendix B13
BS 5446-2 1.3
BS 5499-4 3.45
BS 5499-5 Appendix C11
BS 5839-1 3.48, 16.12, Diagram 9.4
BS 5839-3 9.22
BS 5839-6 1.1, 1.4, 1.6, 1.7, 3.21
BS 5867-2 Appendix B13
BS 5906 3.55
BS 7157 4.8
BS 7273 1.15
BS 7273-4 3.92
BS 7974 0.13
BS 8214 Appendix C14
BS 8313 7.28
BS 8414-1 10.3
BS 8414-2 10.3
BS 8519 3.48
BS 9251 2.6, Appendix E3, Appendix E4
BS 9990 14.5, 14.11
BS 9991 3.29
BS 9999 7.20, 15.8, 15.11, 17.5, Diagram 15.1
BS EN 54-7 9.22
BS EN 81-20 15.11
BS EN 81-58 Appendix C1
BS EN 81-72 15.11
BS EN 520 Appendix B9

BS EN 1366-2 Diagram 9.4
BS EN 1634-1 Appendix C1, Appendix C4
BS EN 1634-2 Appendix C1
BS EN 1634-3 Appendix C1, Table C1
BS EN 1993-1-2 Table B4
BS EN 12101-2 Appendix D5
BS EN 12101-3 16.12
BS EN 12101-6 3.54, Table C1
BS EN 12845 Appendix E3, Appendix E5
BS EN 13238 Appendix B8
BS EN 13501-1 page 76, Appendix B6, Appendix B10, Table B1
BS EN 13501-2 Appendix B22, Appendix B25, Appendix C1, Diagram 9.3, Table B3, Table C1
BS EN 13501-3 Appendix B22, Appendix B25, Diagrams 9.3 to 9.4, Table B3
BS EN 13501-4 Appendix B22, Appendix B25, Table B3
BS EN 13501-5 Appendix B16 to B18, Table B2
BS EN 14306 Appendix B9
BS EN 14604 1.2
BS EN 15102 Table 4.1
BS EN 15650 9.20–9.21
BS EN 15725 Appendix B4
BS EN 50200 3.47
BS EN ISO 306 Appendix B11
Building control body page iv
Definition Appendix A
Building dimensions
Measurement methods Diagrams D2 to D6
Building Regulations 2010 pages iii–iv
Buildings of architectural or historical interest 0.10, 2.22
Building work page iii
Bungalows
See Single storey buildings

C

Cables
See Electrical wiring
Canopies 11.12–11.13, Diagram 11.6
Car parks
Air extraction systems 9.10
Covered 3.28, 3.73
Enclosed
Lifts 3.102
Smoke ventilation 3.75
Escape stairs serving 3.73, 3.75

Heat alarms

Multi-storey flats 3.21

Power supply 1.4

Standards 1.3

Heat and smoke outlets

See Smoke outlets

Heat radiation

Discounting 11.1

Height

Definition Appendix A

Measurement methods Diagram D4

Helical stairs and spiral stairs 3.86

High reach appliances

Fire service vehicle access Table 13.1

High risk

See Places of special fire hazard

Hinges

Fire doors Appendix C10

Historic buildings

See Buildings of architectural or historical interest

Hose laying distances 15.7, Diagram 15.3

House conversions

Conversion to flats 6.5–6.7

See also Loft conversions

Hydraulic platforms

See High reach appliances

I

Inclusive design 0.8

Industrial purpose group Table 0.1

Inner rooms

Definition Appendix A

Means of escape through another room 2.11–2.12, 3.7–3.8

Installers

Certification and accreditation page iii

Institutional premises

See Residential (institutional) purpose group

Insulating core panels

Used internally 4.10

Insulation performance Appendix B19, Table B3

Fire doorsets Appendix C8

Floors in loft conversions 5.4

Glazing 3.88

Insulation (thermal)

Cavity walls Diagram 5.3, Diagram 8.2

Effect on fire performance Appendix B2

External walls 10.6

Integrity

Compartment walls 7.14

Fire doorsets Appendix C1, Table C1

Resistance to fire penetration Appendix B19, Table B3

Internal fire spread

Linings

Insulating core panels 4.10

Requirement B2 page 42

Thermoplastic materials 4.12–4.17, Appendix B14

Structure

Loadbearing elements of structure 5.1–5.4, 6.1–6.3

Requirement B3 page 49

Internal linings

Classification 4.1, Table 4.1

Protection of substrate Appendix B14

Requirement B2 page 42

J

Joists

Openings for 5.9, 7.6

Junctions

Compartment wall or floor with other walls 7.12–7.14

Compartment wall with roof 5.11–5.15, 7.15–7.18, Diagram 5.2

Double-skinned insulated roof sheeting Diagram 5.2

Fire-stopping 5.11, 7.15–7.18, Diagram 5.2

K

Kitchens

Extract ductwork 9.15

Heat alarms 3.21

Inner rooms 2.11–2.12, 3.7–3.8

Insulating core panels in 4.10

See also Cooking facilities

L

Ladders

Means of escape 3.85

See also High reach appliances

Landings

See Protected entrance halls/landings

Large dwellinghouses

Fire detection and alarm systems 1.5–1.7

External walls adjacent to 3.63–3.64, 11.10, Diagram 3.10

Multi-storey flats 3.21

Rooflights of thermoplastic materials 4.14

Separation of adjoining 3.70

Thermoplastic lighting diffusers in ceilings 4.16

Use of space below diffusers or rooflights Table 4.2

Use of space within 3.78–3.80

Use of uninsulated glazed elements on escape routes Table B5

Ventilation ducts 9.7

Protective barriers

Flat roof forming escape route 3.30

Pumping appliances

Fire service vehicle access Table 13.1

See also Fire service facilities

Pumps

Sprinkler systems Appendix E6

Purpose groups 0.14–0.17

Classification Table 0.1

Definition Appendix A

Minimum periods of fire resistance by purpose group Table B4

PVC

See uPVC

R

Radiation

See Heat radiation

Raised floor

See Platform floors

Ramps and sloping floors 3.40

Refuse chutes and storage 3.55–3.58

Regulatory compliance page iv

Relevant boundaries 11.4–11.5, Diagrams 11.1 to 11.2

Definition Appendix A

External walls 1000mm or more from relevant boundary 11.9

External walls within 1000mm of relevant boundary 11.8

Separation distances for roofs 12.3–12.9, Tables 12.1 to 12.3

Small residential buildings Diagram 11.7

Residential (institutional) purpose group 1.13, Table 0.1

Residential use

Purpose groups Table 0.1

Revolving doors 3.98

Risk

Insurance 0.7

See also Places of special fire hazard

Rolling shutters 15.12, Appendix C9

Roof coverings

External fire spread 12.1–12.9

Separation distances for roofs 12.3–12.9, Tables 12.1–12.3

Junction of compartment wall with roof 5.11–5.15, 7.15–7.18, Diagram 5.2

Performance classification Table B2

Slates and tiles 5.22

Thatch 12.9

Wood shingles 12.9

Rooflights

Area measurement Diagram D3

Definition Appendix A

Definition of ceilings 4.6

Fire resistance 4.7

Plastic 4.7, 4.14, 5.12, 12.5–12.7

Fire resistance Appendix B12

Layout restrictions Diagrams 4.2 to 4.3

Limitations on spacing and size Diagrams 12.1 to 12.3

Limitations on use 12.5–12.6, Table 4.2, Tables 12.2 to 12.3

Surface area Appendix D3

Unwired glass in 12.8

Roofs

Area measurement Diagram D3

Cavity Diagram 8.3

Elements of structure 5.3, 6.2

Fire resistance Appendix B15 to B18, Table B3

Insulated roof sheeting 5.14

Junction with compartment wall 5.11–5.15, Diagram 5.2

Rooftop plant

Height of top storey in building Diagram D6

Separation distances 11.14, 12.3–12.9, Table 12.1

See also Flat roofs; Pitched roofs

Roof space

Cavity barriers Diagram 2.3

See also Loft conversions

Rooftop plant

Height of top storey in building Diagram D6

Rooms

Definition Appendix A

S

Safety signs and signals
Fire doors Appendix C11
Seals
Proprietary pipe seals 9.3
Security
Compatibility with escape 3.77, 3.91–3.92
Self-closing devices
Common corridors 3.35–3.36
Definition Appendix A
External escape stairs 3.68
Fire doorsets 3.36, Appendix C5 to C6
See also Automatic release mechanisms
Semi-detached houses
B3 requirement page 49, 5.5
Separated parts of buildings
Compartment walls 7.9
Definition Appendix A
Separation distances
Buildings with sprinkler systems 11.21
Canopies 11.12–11.13, Diagram 11.6
Roofs 11.14, 12.3–12.9, Table 12.1
See also Unprotected areas
Service openings
See Openings
Shafts
See Firefighting shafts; Protected shafts
Sheltered housing 0.11, 0.17
Central alarm monitoring 1.12
Definition Appendix A
Fire detection and alarm systems 1.12–1.13
Shop and commercial purpose group Table 0.1
Showers
Inner rooms 2.11, 3.7
Single storey buildings Appendix B26 to B27
Definition Appendix A
Minimum periods of fire resistance Table B4
Site (of a building)
Definition Appendix A
Skirtings
Definition of walls 4.3
Slates and tiles
Fire-stopping of junctions 5.22
Sleeving for pipes 9.5, Diagram 9.2
Sloping floors 3.40

Smoke alarms
Access rooms 2.12, 3.8
Circulation spaces 1.9
Material alterations 1.9
Multi-storey flats 3.21
Power supply 1.4
Standards 1.2
Smoke and fumes
Smoke leakage of fire doors Appendix C1 to C2, Table C1
Smoke control
Common escape routes 3.49–3.54
Compatibility with ventilation and air-conditioning systems 9.11
By pressure differentials 3.54
Smoke detectors
Extract ductwork 9.9
Fire and smoke damper activation 9.14, 9.22
Smoke outlets
Free area of smoke ventilators Appendix D5, Diagram D7
Mechanical 16.11–16.12
Natural 16.5–16.10
Outlet ducts or shafts 16.13–16.14
Provision 16.1–16.12
Space separation
See Separation distances
Specified attachments 10.9–10.15
Definition Appendix A
Spiral stairs 3.86
Sprinkler systems Appendix E
Basement storeys 16.11
As a compensatory feature Appendix E4
Dwellinghouses with floor more than 4.5m above ground level 2.6
Effect on boundary distance and permitted unprotected area 11.21
Effect on hose laying distance 15.7, Diagram 15.3
Effect on minimum periods of fire resistance Table B4
Loft conversions 2.23
Multi-storey flats 3.21, 7.4
Portal frames 11.15
Provision Appendix E2 to E4
Water supplies and pumps Appendix E5 to E6

Stability
B3 requirement page 49
See also Integrity

Stairs
Width measurement Appendix D4
See also Common stairs; Escape stairs; Protected stairways

Standard fire tests
See Fire performance

Standby power supplies 1.4

Storage and other non-residential purpose group Table 0.1

Storey exits 2.13, 3.10, 3.30, 3.35, Diagram 2.5
Definition Appendix A

Storeys
Definition Appendix A
Height of top storey Diagram D6
Number Diagram D5

Stretched-skin ceilings
Thermoplastic material 4.17

Structural frames
Fire resistance Table B3

Structural loadbearing elements
See Elements of structure

Student accommodation
Fire detection and alarm systems 1.11

Stud walls
Cavity barriers 5.21
See also Partitions

Substrates
Fire tests Appendix B8
Lining to wall or ceiling Appendix B14
Roof coverings 5.13, 7.16, Diagram 5.2

Suppliers
Accreditation page iii

Suspended ceilings
Definition Appendix A
Insulating core panels 4.10
Lighting diffusers 4.15–4.16, Appendix B12, Diagrams 4.1 to 4.3, Table 4.2
Thermoplastic material 4.17

T

Tension-membrane roofs and structures 4.9

Terrace houses
B3 requirement page 49, 5.5

Thatch 12.9

Thermal radiation
See Heat radiation

Thermoplastic materials
Classification Appendix B13
Definition Appendix A
Fire performance Appendix B11 to B14
Glazing 4.13
Insulated roof sheeting 5.14
Lighting diffusers
  Forming part of a ceiling 4.15–4.16, Diagram 4.1
  Layout restrictions Diagrams 4.2 to 4.3
  Suspended ceilings 4.17, Appendix B12, Table 4.2
Rooflights 4.14, 5.12
  Layout restrictions Diagrams 4.2 to 4.3
  Limitations on spacing and size Diagram 12.1
  Limitations on use 12.6, Table 4.2, Table 12.3
Suspended or stretched-skin ceiling 4.17
Windows 4.13

Tiles and slates
Fire-stopping of junctions 5.22

Toilets 2.11, 3.7
Escape lighting 3.42

Trap doors
Definition of ceiling 4.6

Travel distance
Common corridors Diagrams 3.7 to 3.8, Table 3.1
Definition Appendix A
Measurement methods Appendix D3
Single storey flats 3.18, Diagram 3.3

Turning circles
Fire service vehicle access route specification 13.4, Diagram 13.1, Table 13.1

Turntable ladders
See High reach appliances

U

Unprotected areas
Areas disregarded in assessing separation distance 11.11, Diagram 11.5
Calculation of acceptable unprotected area 11.16–11.20, Table 11.1
  Effect of sprinkler systems 11.21
Definition 11.6–11.7, Appendix A
External wall 1000mm or more from relevant boundary 11.9
  Permitted unprotected areas 11.11, Diagram 11.7, Table 11.1
External walls of protected stairways 11.10
External wall within 1000mm of relevant boundary 11.8
Fire resistance 11.6–11.7, Diagram 11.4
Small unprotected areas 11.11, Diagram 11.5
Space separation Diagram 11.1

uPVC
Rooflights 12.7, Tables 12.1 to 12.3
TP(a) rigid sheet Appendix B13

Utility rooms
Inner rooms 2.11, 3.7

V

Vehicle access
See Fire service facilities

Ventilation
Protected shafts 7.28
Smoke control 3.49–3.54

Ventilation ducts 9.6–9.22
Appliances 9.23
Openings in compartment walls or floors 7.20
Passing through fire-separating elements 9.12–9.22
Protected escape routes 9.16–9.17, Diagrams 9.3 to 9.4
Protected shafts 7.26
Protected stairways and entrance halls 2.9, 3.23

Vertical escape
See Escape stairs; Protected stairways

Vision panels 3.97

W

Walls
Cavity barriers in stud walls or partitions 5.21
Definition 4.2–4.3
Elements of structure 5.3, 6.2
Fire resistance 8.3, Table B3
See also Cavity walls; Compartment walls; External walls; Internal linings

Water supplies
Firefighting 14.12–14.13
Sprinkler systems Appendix E5 to E6

Water supply pipes
Enclosure Diagram 9.1

Windows
Definition of walls and ceilings 4.2–4.3, 4.5–4.6
Emergency egress 2.10, 3.6
Replacement 2.18–2.20
Thermoplastic materials 4.13
See also Glazing; Rooflights

Wiring
See Electrical wiring

Wood shingles 12.9

Workmanship and materials page iii
Regulation 7 pages 76–77